AF711280

Jungle Cowboy

By the same author

LEEMO

MORE ABOUT LEEMO

Jungle Cowboy

Stanley E. Brock

Illustrated

TAPLINGER PUBLISHING COMPANY
NEW YORK

First published in the United States in 1972 by
TAPLINGER PUBLISHING CO., INC.
New York, New York

Copyright © 1972 by Stanley E. Brock
All rights reserved. Printed in the United States

No part of this publication may be reproduced or transmitted in any form or by any means, electronic or mechanical, including photocopy, recording, or any information storage and retrieval system now known or to be invented, except by a reviewer who wishes to quote brief passages in connection with a review written for inclusion in a magazine, newspaper or broadcast.

Published simultaneously in the Dominion of Canada by
Burns & MacEachern Ltd., Ontario

Library of Congress Catalog Card Number: 79–163885

ISBN 0–8008–4444–0

Second Printing

CONTENTS

	Acknowledgements	6
1	I suddenly meet a new way of life	11
2	The ranch	21
3	First steps of a tenderfoot	29
4	I cross swords with Cang	35
5	Cracked ribs and wildlife	40
6	Round-up	49
7	Some mechanical problems	63
8	Trail drive	74
9	Through the forest	85
10	Brave hounds and Shiwishi-win	101
11	Leemo and Co	116
12	Zero-two-Zulu and the Wai-Wai Indians	131
13	More animal mischief	149
14	More round-up, and the week the army came	166
15	New horizons	178

ACKNOWLEDGEMENTS

My profound thanks are due to Patricia K. Tarnawsky of the *Reader's Digest*, New York, who urged me to write *Jungle Cowboy*, and for the innumerable fine suggestions she made about the manuscript; and also to Lois Leyh for her encouragement and inspiration. Without them *Jungle Cowboy* would never have been written.

ILLUSTRATIONS

between pages 24 and 25

Stanley E. Brock
Some of the country between the Guyana coast and the Dadanawa
The Rupununi River
Wapishanas on their way to the Dadanawa
Wapishana vaqueros
Swimming the herd
The author taking a coffee break during a trail drive
Lassoing a renegade
A thousand head of Longhorns in the Dadanawa

between pages 40 and 41

Longhorns on the trail
Branding time
Swimming the horses
The author with Leemo
A playful bite on the shoulder
Leemo walking in the forest
At times Leemo was strangely uninterested in horses
Leemo was not always in a harmless mood
Leemo by the Rupununi River
Roland, the puma captured as a mate for Leemo
Roland in close-up

between pages 48 and 49

Climbing Shiwishi-Win with Mr Manager
A cattle killer brought to bay
Mr Manager and family
A jaguar

Shadow
A snack at the store-room for Choo-Choo and the Margay
Choo-Choo would never allow the Margay to stray too far from her side
Choo-Choo
Choo-Choo at the store-room door
Nothing of a chewable nature was safe from Choo-Choo
There were times when Chico could be dangerous

between pages 80 and 81

A giant armadillo drags the author
A tapir caught in a heavy game net
Tiny, the tapir
Big Jacko and Tiny were the best of friends
Irene with Tiny and the two young tapirs
A prehensile-tailed porcupine

between pages 96 and 97

A Tamandua carrying its young
A giant ant bear carrying her cub saddled on her back (bottom picture)
The very rare pygmy anteater
A capybara, the world's largest rodent
A sting ray speared in the New River
Two giant otters inspect a basking caiman
Piranha

between pages 120 and 121

A mussurana swallowing a rattlesnake
The author struggles with the murderous coils of an anaconda in the Rupununi River
The author and Jezebel
Jezebel coming in for a landing (photograph by Ralph Nelson II)
We transported Jezebel in the back of a Land-Rover
Jezebel at rest

A king vulture
A Wai-Wai girl with a toucan
The Wai-Wais give Zero-Two-Zulu a careful inspection
Shooting the rapids in the Essequibo

between pages 128 and 129

The Wai-Wai hunting dogs are tethered on a raised platform
Painting arrow heads with curare
A Wai-Wai draws his 8-foot bow
A successful Wai-Wai hunter returns with a tapir
The secret of sleeping in a hammock is to lie across it
Wai-Wai women spinning cotton

between pages 144 and 145

A Wai-Wai woman making a cassava grater
Making farine from bitter cassava
Trio Indians
Trio women along the trail to the Paroe River
A Trio Indian along the trail to the Paroe River

To

Jimmy Brown,

the greatest outdoorsman I have ever met,

and

Irene Fraser,

a beautiful lady

I

I SUDDENLY MEET A NEW WAY OF LIFE

There is only one town in Guyana of any size, the capital Georgetown. It sits on a mud flat at the mouth of the Demerara River facing a murky Caribbean permanently discoloured by the wash of the Essequibo and Orinoco deltas only a few miles to the north. It's a bustling city of wooden buildings and narrow streets, choked with cars, donkey carts, horse-drawn wagons, street vendors' hand barrows and thousands of bicycles. A melting pot for the races of the world; African, Indian, Portuguese, Chinese, European, Amerindian, and their various mixtures, all struggling with each other for possession of highway and sidewalk.

I found myself there in August 1953, at the end of a three-week ocean voyage from the port of Southampton. I was 17 years old, stepping out alone into the world for the first time, with nothing but £10 in my pocket and a thirst for adventure. Barely a month before I had turned my back on the medieval walls of one of Britain's older public schools. For five years of my life I had been strangled by a stiff white collar, black tie, grey drainpipe trousers and an ill-fitting jacket. I had been herded like a convict, carrying armfuls of Chaucer, Homer, Henry VIII and the Holy Bible, between buildings called The Tower, Beaufort, MacNutt and John O'Gaunt's Kitchen. The fact that rebellious barons had sat between those same dank walls and plotted the demise of everyone from their king to the buxom maid who passed by the tinted windows on her way to the millstream for water, held not the slightest interest for me. I was a prisoner of the establishment, sent there by my elders to be beaten by stern schoolmastes and enslaved by older, privileged boys. I was called a 'curator' and between the ages of 12 and 15 was subject to a 19-year-old budding baronet, who required me to clean his shoes, brew his tea and write his history essays. The only good thing I can remember about my life as a schoolboy was the opportunity, once each year, to thrash some of the aristocratic little bastards during the boxing championships.

Yet, perhaps I judge my scholastic days unfairly, for the institution was set in the beautiful countryside of southern England. It was there that I developed a love for gurgling streams, leaping salmon, birds, butterflies, bats, squirrels, the elusive stoat and the artful fox. I spent every available hour sitting quietly along the nearby river bank watching nature pass undisturbed about its business. It turned my imagination towards South America and the mighty wildlife reservoir that lay between her shores. My father had worked there once with the colonial government, and his tales of its boundless animal wonders were legion. He was a telephone engineer and had been sent there to supervise the installation of a new communications system. He was a man of vision and realized that the business world of Britain was demanding higher and higher academic qualifications of their young employees. My reports were far from giving the impression that I would ever attain much distinction, and it loomed painfully obvious that on the ladder of a highly competitive society I was destined to remain on the bottom rung.

Therefore, my father decided to test the mettle of my own resources and send me to Guyana. It was called British Guiana in those days before independence. The only one of Her Majesty's colonies on the continent of South America, wedged between Venezuela and Suriname on the north-east coast, and pushing south for 83,000 square miles to the Brazilian border, within two degrees of the Equator. It was a land of unexplored interior and opportunity, and a good place to send a youngster who had not done well at school.

While walking through this quaint and colourful metropolis of Georgetown, I chanced upon a slate above a doorway which read:

Rupununi Development Co., Ltd.
Head Office of the Dadanawa Ranch.

It had a dingy entrance off a dirty pavement on a narrow side street. A shaky stairway led to a wooden door that had once, very long ago, been painted light brown. I grasped the door knob and turned, opening almost by accident the most exciting chapter of my life.

The office was empty except for a coloured clerk who sat at a desk by the window. A breeze blowing in from the harbour scattered papers in a disorganized fashion about his desk. He looked at me almost in disbelief when I enquired about a job.

"At the ranch?" he asked.

"Yes, at the ranch," I replied.

"Well, it's run almost entirely by Indians," he told me. "It's in the Rupununi River district, 350 miles from here on the Brazilian border. But we do need someone up there who can read and write. If you like, we'll give you a try."

I indicated that I would be ready to travel as soon as possible, and asked what the duties entailed. This question rather baffled the clerk who admitted that he had never been to the ranch.

"We send a radio message when we want cattle driven to market," he said. "You'll find a Scotsman up there, name of Jimmy Brown, he'll tell you what its all about. The pay is 40 dollars a month and keep. A plane flies up every two weeks with supplies. The next one's on Thursday."

In answer to further questions I learned that the ranch was owned by a public company and the land held on a ninety-nine year lease from the Crown. Most of the shareholders lived abroad, but were represented locally by a small board of directors, some of whom held no shares at all, but merely acted as legal representatives. The ranch paid excellent dividends, and it seemed to be run on a skeleton staff left largely to their own devices. Jimmy Brown was the construction manager and engineer in charge of building fences and corrals. Apparently he was not responsible for the cattle which were handled by the Indians, and any enterprising youngster that the clerk could find in Georgetown to ship up there. He gave me the impression that his previous efforts in this department were only mildly successful. Three applicants had been sent to Dadanawa during the last nine months. None had stayed.

I had not the vaguest idea what to expect when I arrived there, but I could see from the map that the ranch was in the remote north-eastern part of the Amazon basin. Humming birds, toucans, alligators, giant snakes, cannibal fish, and jungle cats were already flying, swimming, slithering and stalking through the theatre of my mind.

The war surplus Dakota groaned off the runway and struggled for altitude. The last signs of the airport slid quickly by the narrow windows as we banked steeply over the Demerara River and headed south. For a minute I could see small farms, clearings, banana plantations and sugar cane, fragile wooden shanties and palm roofed shacks. Then nothing, barely two minutes from the airport and the threads of civilization were suddenly gone. We were over a carpet of green, stretching like a sea in every direction, clothed in a thin film of mist, so low, that here and there the larger trees stood well above it. Not beautiful, not full of colour, just green—light green and dark green; and nowhere, no matter how hard I strained, could I see the forest floor through the canopy of leaves. This was an ocean of jungle that, if a straight line were drawn south to bypass the Rupununi Savannahs, it would stretch for 1,700 miles broken only by rivers. The biggest forest in the world, and the least explored.

After forty minutes the terrain became progressively more rugged. Soon we were nosing our way up through banks of cumulus to gain altitude, making the old bomber pitch and roll. Condensation clouded the windows and the temperature in the cabin rapidly dropped. I was one of two passengers, sitting on a metal seat with a steel spar and bulkhead as a backrest. My companion was clearly Indian, barefooted, copper-skinned, with striking Mongolian features, and clad in rough khaki shirt and trousers. His head had been recently shaved, obviously due to an injury, for his scalp and forehead were terribly scarred. I did not wish to stare at the man nor ask him the cause of his misfortune. He sat motionless and uncommunicative staring at the floor between his knees.

We were surrounded by tons of cargo lashed to the bare deck with coarse rope. There were heaps of barbed wire and sacks of flour, one of which had become snagged and the contents were being gradually evacuated in an ever-widening pool. I got up and moved it, wedging it between two 40-gallon drums marked "KEROSENE". There were bales of cloth, and dozens of wooden boxes which said, "AMMUNITION, 20 GAUGE CARTRIDGES", or "GUNPOWDER, HANDLE WITH CARE"; axes, saws, pots and pans, sacks of sugar, rice, iron bars, rods and galvanized pipe. Standing amidst all this paraphernalia was a very depressed looking black-and white calf. It rocked with the listing of the aircraft, feet splayed well apart for balance. A huge rope tied in a bowline around its neck tethered the unfortunate creature to a sturdy spar. It seemed very young to be without its mother and about every five minutes gave a mournful "muu-uuh".

At 10,000 feet we broke the ceiling and settled down to the methodical drone of cruising engines. Above, the heavens yawned in a cerulean arch, the brilliant blue splattered here and there with the probing tentacles of cirrus. Ten minutes later the cumulus layers began to break up and once more the arboreal wilderness lay mapped out below us. Suddenly, the boundless trees were split by a mighty wandering river nearly a mile wide, a network of ox-bow lakes and foam-flecked raging rapids. I knew from the map that this must be the Essequibo, largest river in Guyana, draining the country north and south for 620 miles. Soon it was behind us and we were into the mountains, but the jungle still fought for control. Only where black granite walls plunged in craggy palisades did the trees fear to tread.

I had been told that the flight to the savannahs would take an hour and forty minutes, and we had been airborne for well over an hour. I pressed my face eagerly against the window and squinted southwards.

Sure enough below the horizon the sea of green washed on an ever-widening pale-yellow shore.

The landing was rougher than I'd anticipated. However, the space allocated for the arrival was obviously sparse and the technique more than justified. The pilot dropped her in from the second floor with a jar that should have made the wheels come up through the wings. I lurched against the seat belt. Then he jumped on the brakes, and I rocked back and hit my head on the bulkhead as the machine vibrated like a plucked guitar string. The entire cargo slid forward a good 6 feet and the sad calf sank to its knees with a groan of despair. Through the whole performance the Indian remained completely unmoved. His attitude reassured me. I was now convinced that this was a normal routine landing, and not an emergency.

A member of the crew made his way aft from the cockpit, scrambling on hands and knees over the cargo. He wrestled with a series of levers on the massive door and swung it open before the aircraft had rolled to a complete stop. A blast of hot air rushed into the cabin as the roar of the engines died away.

"This is Wichabai airstrip," the crewman said. "If you want Dadanawa Ranch you get off here."

I climbed out, dragging my single suitcase with me, feeling a little ridiculous standing in the middle of nowhere with a suitcase. Furthermore, the conspicuous labels, "SOUTHAMPTON", "TRINIDAD" and "GEORGETOWN", clearly branded me a city boy, out of his environment.

A dozen Indians were already engaged in discharging the cargo, while a slightly built white man wearing khaki clothes and flowing white beard made his way towards me.

"How do you do," he said in impeccable English. "My name's McKenna. I'm a Roman Catholic missionary working with the Wapishana Indians." He extended his hand which I shook warmly, thankful to have found a friend so readily.

"I'm Stanley Brock," I replied, "I've come to work at the ranch."

"Oh, then you'll be looking for Jimmy Brown. He should be here in a minute—may have had some trouble getting his Land-Rover across the creek. We've had a lot of rain and the river's up. Ah, there's my man!"

He motioned towards the bald-headed Indian now descending from the aircraft. "Been to the hospital. Got himself mauled by a jaguar, lucky to be alive. . . . Well, here comes Mr. Brown now!"

I turned to see a battered, mudsplattered Land-Rover come bouncing down the trail towards us. It had no top and no doors and the windshield was cracked. The whole vehicle was enveloped in a cloud of black smoke. At a guess, only two cylinders were firing and the exhaust was severely cracked or just not there. The noise was more than sufficient to drown the sound of the departing aircraft as it taxied for take off.

Mr. Brown switched off the ignition as he passed and put the car into a U-turn by the pile of freight. There was a scream of grinding, slipping cogs as he rammed it into first gear, and shuddered to a halt within a foot or so of the pile. Jimmy Brown was obviously a master in the art of stopping a car without the use of brake linings. He slid from his seat without even a glance in our direction and began heaving sacks of flour into the open back.

"Mr. Brown is a man of very few words," said the padre. Then added with emphasis as he moved away, "But a good one."

I walked over to Jimmy, who had just picked up the leaky sack of flour and apparently not noticed the stream of contents that immediately made exit from one corner.

"That one's got a hole in it," I volunteered.

Jimmy paused for a moment with sack at belt level. He glared at me through gold-rimmed spectacles under the brim of a shapeless felt hat. His face was weathered like ancient parchment. The blackened end of a well-chewed cigarette that had not burned for some time hung from his lips. He said nothing, but flung the sack with great force into the Land-Rover and picked up another.

"My name's Brock," I went on. "I've been sent up here to work."

Mr. Brown stopped again for a moment, looked at me and grunted. I thought it best not to extend my hand in the traditional manner. It seemed unlikely that he would be interested in shaking it, and most probably it would have been left hanging in mid air, alone and rejected. He stared at me through a long silence, removed the homemade cigarette from his mouth and examined the charred end. Then he spoke for the first time, with a clipped but unmistakable Scottish accent.

"Fer Christ's sake! Yer only a kid. I ain't got time to look after you."

"I promise not to be any bother Mr. Brown," I interrupted hastily, "and I'm not afraid of hard work. Can I help you load up the Land-Rover?"

"No," he barked. "It's finished. Get in and let's go—but don't expect me to look after you, and don't come yellin' when some old cow sticks 'er 'orn up yer arse, or an 'orse kicks yer in the balls."

He pressed the starter and the engine turned over weakly. He pressed it again with a similar response. Mr. Brown mumbled something under his breath, then punched the starter button three times with the heel of his hand and violently kicked the accelerator. The engine wheezed, coughed and then abruptly died.

"Goddam son of a bitch," he said, "needs a push."

Jimmy looked over his shoulder and fired a staccato of foreign words at a group of Indians who still stood close by with the bald-headed jaguar hunter. They came running over at the command, seized the vehicle, and with great shouts of delight began to propel it furiously down a grassy slope. I leapt out to help in the hopes of impressing Mr. Brown, stumbled and fell, in the process tripping up three of the Indians. We sprawled on top of each other, the Indians roaring with laughter. By this time the jeep was 30 yards ahead and the engine roared to life in a cloud of smoke. I scrambled to my feet and sprinted after it. Jimmy, with no thought of waiting, changed into second gear. The sacks of flour in the back were my target. I flung myself frantically amongst them and held on grimly. I groped for a solid hold and found the anchored spare wheel. We were at 40 miles per hour and hurtling along a sandy savannah trail when I finally hauled myself into the front seat gasping for breath.

Fifteen minutes later I stood on the edge of the Rupununi River, a little bruised and extremely dirty. This was obviously the end of the road, although it would not have surprised me if Mr. Brown had driven the Land-Rover into it and ferried us across by sheer will power and stomach muscles.

Brown muddy water raced level with the banks, surging savagely around strained tree roots, sending shock waves through their branches, whipping them like misused fly rods. Logs and dead limbs cut adrift by the rising water were carried rolling and tumbling downstream, occasionally their shape indistinguishable from the gnarled black heads of the watchful caiman who lurked expectantly in quiet eddies sheltered by massive tree trunks. Islands of 'fire' ants floated helplessly by—red, crawling, heaving, undulating ant-bergs, some 3 feet high and 6 feet wide; millions of desperately struggling insects, their grappling bodies holding them together in a living, swimming life raft.

A pair of red-headed vultures circled with scarcely a wing beat in an invisible thermal. A white-tailed hawk stood sentry in the lofty branch of a nearby mora. There would soon be food, a drowning fledgling or armadillo, a fleeing mouse; the angry river was bound to cast up its toll.

The continuation of the trail could be clearly seen on the other bank. No doubt, in dry weather, the water level dropped sufficiently to allow a vehicle to cross. But now, as there was no boat in sight, I wondered just how we were going to do it. At this point Mr. Brown took an automatic shotgun from the car and fired four shots rapidly into the air. I took the action to be a signal, but not wanting to burden him with any unnecessary conversation, decided not to ask but wait and see. By this time I was engulfed in a dense cloud of tiny black flies. The exposed skin on my face, neck and arms began to burn, and I quickly rolled my shirt sleeves down and turned the collar up for protection. Each bite was like a needle prick and oozed an equivalent drop of blood. The flies were no larger than a pinhead at first, but rapidly swelled like balloons to twice their size as they filled with blood. They were sluggish, and a swat of the palm killed them by the hundred. In a few minutes I looked as though I was suffering from a new disease called the "bleeding measles".

"Kaboura flies," Jimmy commented with a trace of a smile. "They love new white skin."

He was covered in them, but seemed not to notice and busied himself with several futile attempts to relight his black tobacco. If he ever could get the thing alight, I thought, it would make an excellent smoke screen.

A few minutes later a dugout canoe swung into view, paddled fiercely by three Indians. It approached the bank with incredible speed but a second before impact the men thrust their paddles deep into the water and the bow nudged gently ashore. The boat was about 30 feet long and must have been carved from an immense tree, although the beam was narrow, perhaps 4 feet at the most. This made the craft extremely unstable and rocked wildly as soon as I stepped aboard. Hastily I dropped my suitcase in the bottom and sat down. Four sacks of flour were loaded and by the time we were fully manned with Jimmy astride one of the sacks, gun across knees, water was perilously near the gunwales. As we pushed clear of the shelter of the bank I realized the immense power of the current. It caught the bow and whipped it viciously downstream. Desperately the paddlers lent their combined efforts and slowly brought her around. Then together plowed their blades deep and we inched our way upstream.

After fifteen minutes we had made good only 100 yards and crossed to the opposite bank. It seemed heavily wooded and impenetrable, until suddenly a tunnel appeared where the brush had been cut. Skilfully the bowman nosed the boat into it and we were immediately in quieter water. The sweat-soaked men relaxed and moved the canoe gently forward with only occasional strokes.

In places we had to duck to avoid low moss-covered branches. There was an eerie climate about the tunnel. The sky was obscured by overhanging foliage and it conjured up thoughts of snakes falling from trees or striking at a face as we passed. By this time I had become more accustomed to the balance of the boat, acquired a spare paddle and began to help. All at once the tunnel ended and we glided into a wide, still lagoon. Five more minutes of gentle paddling elapsed. The forest became thinner along the bank, gradually changing into open plain. We turned a bend and there, a stone's throw ahead, lay a cluster of white-painted wooden buildings, neatly arranged on stilts. The Dadanawa Ranch House. I had arrived.

II

THE RANCH

The houses stood on a small hill about 100 yards from the lagoon. Six buildings rambling over several acres overlooked by a range of mountains barely 3 miles to the north.

I struggled with my heavy suitcase towards the nearest house. Jimmy Brown had strode ahead as soon as the boat docked and already disappeared. It was obvious that I would be left to make my own self at home and decided to make the most of it. I pushed my way clumsily through a multitude of ducks, turkeys, fowl and baby chicks, who quacked, gobbled, and clucked between my feet as if I were carrying their morning feed in a bucket instead of my worldly possessions. The house was surrounded by a low wire fence, and the entrance, a small green gate, was blocked by an enormous red recumbent sow, saddled, straddled, snuggled and suckled by at least a dozen grunting piglets. I continued my approach, assuming that her ladyship would rise and move aside. After all, what more inoffensive an animal than a farmyard pig?

Perhaps some people are lucky enough to read about the temperament of a motherly sow, perhaps others are forewarned. Others, no doubt, are destined to learn everything the hard way, a method which has the advantage of conveying a message that sticks. On that day I became convinced that I belonged to the latter category.

I was 10 feet from the old girl when she came to her feet. Not the swaying, flabby movement that one would expect from her species, neither did her beady eyes telegraph the intent. In a flash she had four legs under her and with a snort came for me like a mad dog. The suitcase saved me. I dropped it in her path and fled for the fence. She collided with the bag, and then swerved to get me. Five feet of wire mesh fence barred my way. There was no time to gauge a jump. Teeth were clanging shut like spring-loaded steel traps a yard behind me. I had one-fifth of a second to make up my mind. Either look like a damn fool in a heap on the far side of the fence, or a bleeding idiot if I refused to jump. I flung myself upwards, head first with arms outstretched. The wire caught me across the thighs and pitched me violently face downwards towards the ground. I felt the wire bite into my skin above the knees. It held for a fraction, and then with a sound of tearing khaki cloth, I sank to earth in an elegant crumpled pile.

I rolled over and looked back. The wire separated me from a pink, frothy, cavernous mouth, armed with the meanest array of dental equipment I had even seen. Then the sow turned and lumbered towards the fallen suitcase. She grabbed the handle in her teeth and with a jerk of her powerful neck, wrenched it sideways. To my horror the lid flew open, and a second jerk scattered the contents across the grass. I jumped to my feet with a shout and raced for the gate. In doing so I almost collided with a large negro lady, dressed in neat blue print, white cap and apron, who was screaming at the top of her voice and brandishing in one hand a broom, and the other a bucket of water. I stopped in my tracks as she burst through the gate. She pulled up a broom's length from the sow, now completely absorbed in rummaging through my underwear, and with one swing swamped pig and belongings with the contents of the bucket. Even before the animal could think to retaliate, it was being belaboured furiously on the stern by the business end of a broom, and in a second retreated, vanquished and confused, in pursuit of her brood.

"Good afternoon suh," the negro lady said, turning towards me with a broad benevolent smile sparkling across a face of forty years or so.

"Good afternoon," I answered, still half stunned, and now conscious of torn trousers and blood spattered knees. "Very kind of you to rescue my gear, thank you."

"That old sow's chewed up your drawers a little bit," she observed, displaying aloft, without a trace of embarrassment, a tattered, grubby pair that she had salvaged. "But I guess they're still wearable!"

Together we retrieved my scattered possessions and stuffed them back into the suitcase. I rammed the lid shut, and then said, "I've come here to work, have you any idea where I should put my things?"

"Well the last man stayed right here in this house," she

retorted, waving her hand at the nearest one. "Just climb the front stairway and go through to the room at the back. You'll find a hammock in there. My name's Irene Fraser. I do the cleaning, and there's a cook you'll be meeting. His name's London."

"Thank you very much," I replied. "Tell me, what happened to the last man?"

"Dunno," she returned. "Only stayed two weeks, came on one plane, left on the next. There's been lots like him, isn't everybody's idea of home—Dadanawa!"

Such were my first impressions of the ranch—harsh, cruel, unpredictable, yet set in a beautiful landscape of forest-covered mountains, awesome cliffs, and mile after lonely mile of desolate grassland. It was the home of 30,000 Longhorn cattle, the range bull and wild horse; hunting ground of the Wapishana Indian, jaguar, puma, and the prey they relentlessly sought. The ranch was part of an area of 5,000 square miles of rugged savannah, largely drained by the giant arteries of the Amazon system. These grasslands lay in a north-and-south pattern, bisected roughly by a range of mountains called the Kanakus, running east and west. The Rupununi River cut a gorge through the highland, and from this winding host of rapid and whirlpool, the savannahs derived their name. Dadanawa occupied 3,000 square miles in the southern section of the territory, and in area was one of the largest ranches in the world.

The house that Irene had guided me towards was a rambling construction on tall stilts. It had been painted white, obviously many years earlier, for now the undercoat shone through in large patches blistered by sun, wind and rain. Most of the area between the stilts had been walled in with thick boards, and the sandy soil covered with a layer of rough concrete. In design it would have made an excellent garage or even a hangar for a small airplane, but instead it was the dining-room. The floor above it was held up by timbers of enormous proportions. There were eight, each made from tree trunks of such girth that I could barely get my arms around them. They had a marble-like surface that was roughly hewn into shape with the blade of an axe. Whoever put them there expected them to stand for centuries, and I learned later that the material was known as 'bullet-wood', which is a species of balata, a latex-producing tree akin to rubber.

After climbing upstairs I found that the enduring features of the ground floor were sheltered by a very shaky, shingle clad roof. In fact, there was enough sky showing through that it would have been no great trick to have given a weather report without going outside or looking through a window. I use the word 'window' somewhat advisedly. Glass was non-existent, but there were large openings covered in fine wire-mesh screens. Obviously the intention was to keep the ravenous kaboura flies at bay, an objective totally defeated by the holes in the roof.

By this time my eyes had become accustomed to the darker light of the interior, and my ears tuned to a unique audio of high-pitched squeaking and buzzing. The cedar rafters and precariously attached shingles now commanded a more intense inspection. The roof was some 20 feet above the floor level, and suddenly, to my alarm, I realized that every available square foot was occupied by dripping, dangling bats, and the honey-comb-shaped nest of the vindictive, tropical hornet!

There was an additional outdoor stairway that provided access from the yard to the upper level, should one choose not to enter via the lower dining-room. This also led to an elevated veranda at the head of the stairs which allowed a magnificent view of the nearby Kanaku Mountains. Other than that the house was divided into only two rooms; the front lounge, and my bedroom, which was at the rear and confined by a partition about 7 feet in height. It was bare except for a cotton hammock hanging in one corner. The furnishing in the lounge, however, was the building's most appealing feature. There were two homemade wooden chairs covered in jaguar skin. Each one had been carefully sculptured to fit a man's back in a reclining attitude, and the two arm rests extended straight out for about 40 inches. I sat in one, laid back, and lifted my legs onto the rests. It was fabulously comfortable, and gave a wonderful view of the bat and hornet activity in the roof.

After enjoying the chair for a few minutes I decided to go in to the bedroom and experiment with the hammock. It was made of loosely woven cotton strands, beautifully worked, obviously by hand. At each end a length of rope tethered it securely to sturdy iron rings screwed into the wall. I eased myself gently into it, the

Stanley E. Brock (photograph by Henry Herr Gill of the *Chicago Daily News*)

(*left*) Some of the country between the Guyana coast and the Dadanawa. (*below*) The Rupununi River

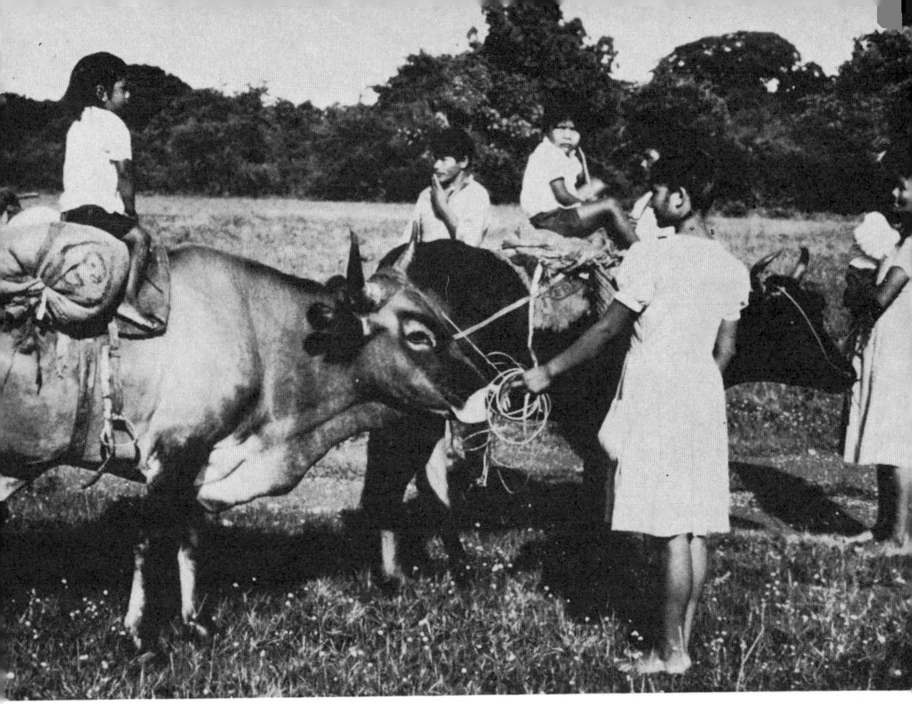

(*top*) Wapishanas on their way to the Dadanawa

(*bottom*) Wapishana vaqueros

(*overleaf*) Swimming the herd

(*left*) The author taking a coffee break during a trail drive

(*above*) Lassoing a renegade

(*overleaf*) A thousand head of Longhorns in the Dadanawa corral

cotton stretched with my weight and settled into a gentle curve about 18 inches above the floor. I was lying down the centre line and in a position not unlike I'd been earlier in the uniquely constructed chair. My feet were in line with my head, but my rear end was about 3 feet lower. Such a V-shaped position would be hard to take in large doses. When I rolled over on my stomach it felt like a back exercise for a ballerina. I got up and tried with my feet at the other end. No good, it was just the same. I was ready to give up and think about making a bed on the floor when I heard Irene's footsteps approaching.

"Everything all right, suh?" she inquired politely.

"Fine, thank you," I replied. "I was trying out the hammock. Very nice, but it looks as if it'll take a bit of getting used to."

"Did you lie across it?" she asked.

"Across it? Well, I laid in it. How do you lies across it?" I questioned.

Irene walked over, opened it with two hands like a newspaper, sat on the edge and rolled her 200 pounds gently and gracefully in. I then realized that the hammock was rectangular in shape and almost as broad as it was long. To lie across meant simply to lie along the diagonal of the square, with head at one corner and feet at the other. The body then lies in a straight line, with no curve in the spine regardless of whether you lie on your stomach, back or side. I tried it and was comfortable at once. The South American Indians invented the hammock, and I was ready to agree, they had a good idea.

Half an hour later a bell rang—a signal which Irene had previously warned would herald all meals. I made my way downstairs and found Mr. Brown already there. He was putting the finishing touches to a hand-tailored cigarette and poring over a two-week-old edition of the Georgetown newspaper, one of fourteen back numbers he'd received on today's plane. Presumably by reading one a day he would get the feeling of being abreast of the news, albeit half a month behind the rest of the world.

The door opened and a black man of large build, dressed in a uniform of white shirt, shorts and apron, walked in. I gathered he was the cook, London, whom Irene had mentioned. He carried an enormous round of roast beef on a platter which he put down on a small side table. I said hullo; London murmured a reply, left the room, and returned moments later with two bowls. One contained potatoes, and the other a curious grey substance, which at first glance looked rather like a cross between grain and grit.

Mr. Brown put down his newspaper and yet unlighted cigarette, got up and walked to the side table. He brandished a large carving knife, stroked the blade several times against a fork and carved himself a generous portion of meat. He then looked disdainfully at the boiled potatoes, grunted, and instead served himself four large spoonfuls of grit.

"You'd better help yourself," he said, sitting down again at the table. "No one else will."

I cut three slices of meat, ragged and ineptly against the grain, trying when I'd finished to straighten the rough edges, lest Mr. Brown should be offended at my inability to carve like a man should. I selected three of the scorned potatoes, and so as not to appear unsociable, a small spoonful of the unappetizing grit. Jimmy Brown was already well started on his meal, spoon in one hand, newspaper in the other. As I sat down he was just disposing of a large mouthful of the grainy stuff, which cracked and ground as if he were chewing a handful of small stones. I ate all the meat and two potatoes and then became inevitably faced with the grit. Cautiously I sampled a small spoonful. It was like eating sand, only it had a slight woody, fibrous texture to it, and virtually tasteless.

"How'd you like the farine, boy?" Mr. Brown asked with a sly grin.

"Is that what it's called?" I asked, making an effort to swallow some.

"Yeah, staple food of the Indians. Don't know why they bother to send up these damn potatoes, waste of freight on the airplane. Beef and farine's all we've ever 'ad. You'd better get used to it. It's all you'll get out on the range during round-up."

Mr. Brown was finished and got up from his chair.

"What's farine made of?" I asked as he left.

"The Indians make it from bitter cassava," he replied as he went through the door. "It's a poisonous root. If it's not processed right it'll kill you as dead as this post."

He slammed the door, and I looked down at the grit, now named farine, in my plate—I wondered if this had been processed right. Perhaps I wasn't ready for farine just yet, and carefully moved it aside and began to work on the last of the remaining potatoes.

The bunk house at Dadanawa was an adobe brick building about 100 feet long and some 30 wide. The sides were only walled up half way, about 5 feet high, and the saddles and other trappings were either hung from wooden hooks or draped along the top of the walls. The roof was thatched with leaves of the Mauritius palm, which grew in great abundance on the savannahs.

I had been directed there by Irene, who so far had proved to be the only person at the ranch who had time or desire to talk with me. She called it the "barracao", which I later discovered was a Brazilian word meaning the sleeping quarters of the cowboys, or 'vaqueros' as they are called. And the man to look for was Charlie, the Indian foreman.

The wooden door was hooked shut with a loop of twisted rawhide. I unhooked it and walked in. Numerous hammocks were slung in all directions from axe-hewn cross-beams. Other than a rough plank table and two benches there were no furnishings, and the dry mud floor was littered with cow hides, lariats and saddlery equipment in various stages of repair. There were about 7 Indians in the building, all lounging in hammocks. Some appeared to be engaged in leatherwork, one plucked at a guitar, others talked. However, the moment I walked in, whatever they were doing stopped immediately, and seven pairs of eyes watched me, unsmiling and with obvious mistrust.

I took a deep breath and said, "Where's Charlie?"

A full ten seconds elapsed before one of the men got up from a hammock and shambled slowly towards me. He was slightly built, about 5 feet 4 inches in height and might have weighed 110 pounds if he was soaking wet. He was dressed in khaki shirt and trousers, barefooted, and walked with a slight limp. A growth on his left eyelid caused it to droop. His age was difficult to estimate—between 30 and 45 was as close as I could guess.

"Me Charlie," he said, and stretched out his hand.

"Stan Brock," I replied, keeping it as simple as possible, and eagerly pumping his hand in as convincing a show of friendliness as I could muster. "I come here to work."

"You know for ride horse?" he asked.

"A little. I want to learn," I answered.

"Okay, tomorrow we will see. Vaquero horse bad, maybe he break your foot like me. Better you learn good," said Charlie, and with that turned and limped away.

I was rather stunned by the attitude of the Indians. They seemed surly, and I felt sure, that if I had not spoken first and asked for Charlie, they would probably have just stared at me in complete silence until I left. They viewed me as another inept white man sent by the mysterious owners of the ranch to spy on them. I couldn't ride, couldn't rope, didn't know where the cattle were or how to catch them. They were responsible for that and felt embittered by my intrusion. An unexpected chasm had suddenly yawned apart between us. I stood naked and helpless on one side, and the clannish Indian, the only people who could help me cross it, stood aloof on the other.

I swallowed, left the barracao and walked back to my house. Tomorrow the task of bridging the resentment gap would begin.

III

FIRST STEPS OF A TENDERFOOT

The Dadanawa corral was a mammoth rustic construction covering an area of at least 7,500 square yards. Its walls were built of massive hardwood rails, each about 12 feet long and a foot thick, stacked horizontally one above the other to a height of 8 feet. Every 10 feet two huge upright posts, one on either side, clamped the rails together and for extra strength were bound with wire. The main corral occupied about one-third of the area, and adjacent to it were three quite small pens which led through gates to a narrow chute. This chute was obviously used for separating cattle. It was approximately 45 feet in length and made of heavy planks bolted to upright timbers. The exit end was blocked by two verticle rails which were hinged at their base and opened and closed like a scissors. I later learned that a man skilled in their operation could trap a steer by the neck by deftly slamming shut the rails as it passed. Just ahead of the scissors were two swing gates used to separate the cattle into holding corrals on either side. The rate of flow of the cattle was controlled by opening and closing the scissors.

The seven vaqueros I had met in the bunkhouse sat along a top rail of one of the smaller corrals smoking black tobacco and with lariats slung carelessly over their shoulders. In an opposite corner a solid mass of some fifty ponies, jostled, kicked, fought, squealed, shoved and crowded together as far away from the Indians as they could possibly get. They were mostly dirty, ill-kept, shaggy-maned, long-tailed animals with flared nostrils and wild wall eyes. There were greys, blacks, bays, chestnuts, buckskins and palominos. There were horses with crumpled ears, and even one with no ears at all. There were ponies with saddle sores that ran the length of their backs, and stallions with cruel scars doubtless won in duels when honour and harem were at stake.

I walked over to Charlie, wished him good morning and enquired whether Mr. Brown would be coming to the corral to look at the horses.

Charlie removed his cigarette, looked at the end, and screwed his deep brown eyes into a narrow line as the black acrid smoke filtered into them.

"Mr. Brown no ride horse," he said forcefully. "Only Wapishana ride horse. Mr. Brown rides jeep and builds corrals."

It was obvious from the tone of Charlie's voice that he was recommending I should follow in Mr. Brown's footsteps on the construction side and leave the business of rounding up cattle to him and fellow Indians. I wasn't quite sure how to handle the situation, but almost without thinking burst out with an atrocious lie.

"The big boss in Georgetown said he wants me to learn to be a vaquero. He said you were the best vaquero on the ranch and would give me horse and teach me."

Charlie looked at me for a moment thoughtfully, not sure whether to believe me or not. To him the big boss was a mysterious figure, so mysterious in fact that he only visited the ranch about once every two years, flying in by chartered plane and leaving again the same day. Charlie's ego immediately began to inflate at the thought that someone so prominent knew that he was Dadanawa's best vaquero, his fame had at last begun to spread.

"Okay," he said. "You ride Butterball."

The foreman shook out a loop in his rawhide lariat and moved towards the huddling, plunging ponies.

Brown dust quickly enveloped the milling animals until only those in the foreground were recognizable. The closer Charlie walked the more frightened they became, until suddenly they broke from the pack and thundered by us at a frenzied gallop. Charlie swung the loop once around his head and threw as a buckskin passed in front of him. Too late—at that instant a faster moving grey cut across the buckskin's path and ran straight into the noose. Charlie gave a curse as the loop snapped shut and leaned back against it. The jerk whipped the grey into a wild twisted jacknife leap which dragged the Indian as if he were on skis a dozen yards across the sandy corral. The horse stopped and turned to face him, ears pointed well forward and neck arched. Charlie worked his way cautiously up the rope whistling softly. The grey pawed the ground with a front foot like a mad bull and blew air down its nostrils in loud fierce snorts. Gradually the foreman eased up, and reached out gingerly towards its nose with his left hand. The horse began to back away and Charlie followed. Without warning the animal reared up and flashed both front hooves in a lightning one-two at Charlie's head. Had they connected they would have crushed his skull, but the Indian leaped nimbly aside and the murderous weapons slashed wickedly, but harmlessly, by. Coolly, Charlie once again approached, this time more successfully. He ran his hand up the big grey's forehead and gradually worked around under its chin and grasped the loop. Immediately the horse began to back up, but as it did the loop slackened and finally pulled over its head. Free, the grey gave a proud toss of its head and snorting violently trotted away to rejoin the others.

"That was Cang," said Charlie with disgust. "I catch by mistake."

"Could I ride him?" I asked.

"You ride Cang and he kill you," Charlie replied, and turned away.

The foreman did not make a second mistake and a few moments later the yellow-coated Butterball was standing quietly at the end of his rope. He was an attractive horse with a Roman nose, and his black mane and tail were a striking contrast to the buckskin colour. He stood about fourteen hands and his strong broad back never quivered as I ran my hand gently over it.

At last I began to feel the first flood of excitement through my veins. Forgotten for the moment were the tensions between Charlie and myself. I had a horse, and was on the verge of my first adventure, a ride with real vaqueros on to the vast Dadanawa range. Butterball tugged impatiently at the rope. He was the model of man's best friend. His feet and chest indicated the strength and endurance of an Arab and the rugged qualities of his Spanish mustang ancestors; but his eyes were as quiet as a lamb's. I looked quickly across at the other animals and saw the ugly head of the big grey fighting his way furiously through the dust and disorder of heaving horseflesh. Perhaps if I rode Cang I might win acceptance by the vaqueros. Almost subconsciously a challenge was born.

While I held Butterball, each of the Indians lassoed himself a horse, and together we led them down to the barracao. Here I was introduced to the Brazilian saddle. It had a high dome-shaped pommel, but lacked the familiar horn of the American western type. The seat was covered in deerskin, and the flaps and stirrup leathers were similar to those found on an English saddle. Two girths were made from braided leather about 3 inches broad and extremely strong. One was cinched around the horse's chest in the normal manner, while the second was drawn quite far back under his belly. The lasso, made of four strands of plaited rawhide, was coiled and tied to the cantle and anchored to a heavy ring woven into the front girth under the rider's right thigh. Thus, when roping cattle, the horse would take the strain sideways from the girth, instead of from the saddle horn *à la* American style. The bridle consisted of four, braided-rawhide reins, and a long shanked forged iron bit. A single piece of leather held it in place over the horse's ears. In addition, there was a plaited cowhide halter, known as a bosal, under the bridle, and about 20 feet of rawhide halter rope, used to tether the horse. Prior to mounting, this was coiled up and fitted under a loose flap of deerskin on the saddle pommel.

The vaqueros were all ready, and it was time to leave. At this point, I discovered that the stirrup irons were much too narrow to accommodate my boots. They were barely 3 inches wide at the base. I glanced quickly at the Indians already in the saddle and saw that they rode with only the ball of their feet in the stirrup, grasping the iron between the third and fourth toe. Hurriedly, I kicked off my boots, swung rather painfully aboard and together we moved off.

We left at quite a slow pace, and I had no difficulty in keeping up with the others. They were holding their ponies on a tight rein while nudging them with primitive roughly-made iron spurs strapped to bare ankles with leather thongs. The horses danced and reared with necks arched and were soon foaming at the mouth and lathered in sweat. There was much exchange of shouts and laughter between the men as each competed against the others to see whose horse was the champion prancer. Butterball on the other hand picked his own pace and settled down at the rear, head nodding almost sleepily a few inches behind one of the horse's tails.

We rode for about an hour across gently rolling terrain, sandy gullies, and stony hills littered with sharp quartz and giant outcrops of black granite. Here and there we crossed narrow swamps lined by tall, stately Mauritius palms. They were similar in stature to coconuts, 40 and 60 feet tall, and festooned with clusters of brown golf-ball-like fruit. Our ponies ploughed chest-deep through the soft mud and water, and here and there the Indians plucked fallen fruit which floated on the surface. The brown planished skin rubbed off at the touch and left an orange-coloured pith a quarter of an inch thick which covered a hard inner nut. I copied the Indians, chewing the pith and throwing away the nut. It tasted rather like date-flavoured peanut butter.

A short while later Charlie called a halt. Apparently we expected to encounter cattle nearby and everyone dismounted. At this point I witnessed a ritual that I would see performed several times a day from then on, whenever we were out riding. First, each man fully slackened both girths on his saddle. Then the halter rope was paid out to its full length and the pony encouraged to walk slowly in a circle around the vaquero. Some horses walked for only a few paces before stopping, stretching out and then urinating. Others had to be walked for three or four minutes, and there was obvious concern when one failed entirely to answer the call. Charlie instructed me to circle Butterball who, an old hand at the routine, responded immediately, much to the foreman's satisfaction. I was cautioned by Charlie that always, prior to an expected sustained gallop, the horse must be given the opportunity to pass urine. "If not," he said gravely, "maybe you kill horse!"

Back in the saddle we cantered to the crest of a nearby ridge. There, immediately below us, about twenty longhorns had caught our wind and were making off like frightened antelope. Charlie spurred his pony and sped down the rocky hill after them, closely followed by the other Indians. Butterball, without command, accelerated so exuberantly that had I not grabbed frantically at the front of the saddle, I would have been left on the top of the ridge.

Miraculously we were safely down the hill and racing across the flat. The bobbing heads of the longhorns were only a hundred yards ahead and Charlie and the boys were already drawing level and beginning to cut the leaders off. Gradually the bunch slowed to a halt and I caught up, my bare toes already skinned and bleeding from holding grimly on to the stirrups. Suddenly, an old black cow darted out of the herd and raced away passing only a yard or two from Butterball's nose. Quite against my will the old cow pony surged forward and was after her. I leaned forward in the saddle, gathered up the reins and for thirty seconds felt like the professional cowboy at work. Butterball rapidly overhauled the cow. Desperately the renegade skidded round in a hairpin turn to the left. Butterball changed ends and stayed with her, but too late, still poised for straight and level flight, I continued the original course and landed unceremoniously on the back of my neck.

I sat up dazed and looked around. Butterball stood facing me fifteen paces off looking through large brown eyes in obvious disgust. Behind him the Indians sat on their horses and roared with laughter. I tried to laugh too, but felt ashamed and ridiculous. It would take a lot to prove that I could make a vaquero; I had been thrown by the tamest most inoffensive horse on the ranch. Now there was nothing else for it but to ride Cang!

IV

I CROSS SWORDS WITH CANG

During the ensuing months I rode Butterball daily, hardening my feet and rear-end, both of which had shed countless layers of skin. I also practised for three hours every day with the lasso, beginning with stationary targets such as the top of a fence post. As my competence increased I found I could catch with varying accuracy, horses and cattle. Unfortunately, these endeavours were somewhat hindered by the lack of professional coaching. None of the Indians, least of all Charlie, displayed the slightest interest in my efforts. I could only learn by watching them, and then trying to duplicate each manœuvre during practice sessions. I tangled myself up with amazing consistency, and burnt the skin off my hands every time the rope slipped through them unexpectedly. Yet, despite the absence of friendly advice that might have corrected these problems more efficiently, my interest in the art grew daily with proficiency.

However, the over-riding obsession throughout these months had been Cang. The more I thought about him the more I became convinced that my only road to acceptance as a real vaquero lay upon his murderous back. I had watched the Indians ride many wild colts but the big grey had always been by-passed when they roped new ponies in the corral. I learned that his reputation was a fearsome one. The name Cang, in fact, was a Brazilian word meaning 'devil'. He had been sold to the Dadanawa Ranch as a *segunda montado*, a colt that had only been saddled twice. Actually, he turned out to be a 10-year-old range stallion who had been ridden several years before, but killed a man, and caused so much other trouble that he was released with a bunch of wild mares and hadn't seen a saddle since. When Dadanawa ran short of horses after a severe bout of trypanosamiasis, they bought anything with four legs they could get their hands on. The Brazilians across the nearby frontier took full advantage of the situation. Cang was not the only renegade that was slipped in amongst a sale of 50-dollar saddle horses.

The animals were bought on contract from other outfits who had more spare time to round them up than the busy vaqueros at Dadanawa. They merely chased the wild herds that ran freely over the Brazilian range. Hit or miss methods resulted in any number from nil to three hundred finally being tricked into entering a corral. Then all the contractor had to do was lasso and halter break the required number. He wasn't interested in a horse's personal disposition, or how old it was. If it was sleek, and looked strong enough to carry a saddle, he burned on the double H Dadanawa brand, delivered it and collected the money.

Most of them managed to carry a saddle; all took exception to the addition of a rider. The one who disliked it most was Cang—the devil, a flea-bitten grey with an ugly head and ears almost as long as a mule's. He was 700 pounds of kicking, biting, bucking fury.

Cang blew a snort of resentment through his nostrils, and strained against the thick rawhide rope that held him fast to the breaking post. He glared at me through wicked wall-eyes, ears flat against his wild unruly mane. I moved slowly towards him, whistling softly, cautiously inch by inch until I grasped the free end of the halter rope and jerked it loose. With a squeal of rage Cang reared up on his hind legs and slashed at me wickedly with front feet. I leaped back with only the brim of my hat to spare, taking the slack of the rope with me. His weight came down against the post and I snubbed his jaw hard up to it. Then I made the end of the rope fast to another nearby post and we were ready to begin.

I brought my saddle from the barracao and put it down several yards away from him. Once more I edged my way forward, armed with a saddle blanket. He couldn't rear up now as his head was tied firmly against the wood, but he strained back until his belly almost touched the ground, and the plaited rawhide halter bit deeply into his skull. Suddenly he rocketed forward with a terrifying snort, battering his head hard against the post, and finally crashing over with a resounding thud on his side. He lay there thrashing wildly with his fore and hind hooves, digging deep furrows into the hard ground and grinding his teeth together like broken glass. Swiftly I slid a small length of cloth across his angry eyes, and tucked the ends under each side of the halter. Then I jerked free the halter rope to slacken his head, and gave him a smart clout on his rear-end with the horse blanket. Cang, mystified by the sudden darkness, quietly regained his feet.

Whistling softly, I folded the blanket in three, and gently rubbed it against his neck. Gradually I worked it down the withers and across his broad quivering back. Now the saddle. Standing well clear of his near shoulder, while cautiously patting his neck with my left hand, I reached forward holding the saddle in my right, lowering it carefully onto the blanket. The action was followed immediately by a muscular eruption, the saddle and blanket were lost amidst flashing hooves, and Cang once again threw his 700 pounds of fury to the ground.

On the third attempt he stood still long enough for me to catch the ring of the front girth under his belly with a long forked stick, and I cinched it up with the leather thong.

Then the rear girth; with even greater care this time, lest he react violently when he felt it tighten around his belly. This was an important step, as the rear girth would prevent the saddle slipping forward over his shoulders if he should decide to buck or run.

Finally the bridle. I worked the bit between his clenched teeth, slipped the headpiece and reins over his ears, and tied the throat lash tightly and firmly.

So far, so good—I gave the nearside stirrup a strong determined pull, to test his reaction to the weight. Cang answered with a lightning sideways one-two with both hind feet. I'd have to watch those, even with the blindfold still in position. I flapped suddenly at his face with my hat, nothing happened; obviously he couldn't see me. The stage was set for the showdown!

The layout around the barracao was not ideally suited to riding spooky horses. There was plenty of open space further out, but to get there one had to navigate a maze of wooden buildings, trees, and barbed wire fences. Anyone with an atom of common sense would have led his horse well clear of these obstacles before climbing on. But the Indians never did. The hazards seemed to be part of the game, and damn it I was going to play it their way!

Working silently I eased free the halter rope and coiled it up in my left hand, which also took a firm grip on the nearside shank of the bit. I was careful not to move the horse away from the hitching post and he still thought he was tied to it. This made it easier to mount while he was relatively still, but had the added danger of a possible collision with the post if Cang took a deliberate dive forward. On the other hand, if I led him clear he would realize at once that he was free, and then it would be very tricky to get in the saddle.

I reached out with my left foot and hooked my third toe into the stirrup. At the same time I stretched over the front of the saddle and grasped the pommel on the offside with my right hand. Then in one co-ordinated movement, as fast as my body knew how, I leaped into the saddle and rammed my right foot into the other stirrup.

As he felt my weight Cang exploded into a dynamic, twisting power dive, and my leg went numb with pain as it banged into the post. He gave vent to his feelings with a fearful squealing snort and leaped high in the air. His head was lost between his back legs and as he came down all four feet were rigid beneath him. My teeth clanged shut as my jaw collided with my chest and I hung on, jarred almost senseless. Then he rocketed forward in a series of low, plunging crow hops, and his mouth frothed with foam as he hacked at the steel of the bit in a desperate attempt to break free. A wooden building loomed into view, no time to go around, no time to stop, nothing could check this headlong rush now. A path 6 feet wide separated an adjacent outhouse from the main building. A stout pole traversed the gap, festooned with sun-dried meat. Cang charged through. I shielded my face with right arm and the impact shattered the pole with a crack like a rifle shot, showering meat in all directions. We raced across a stretch of open ground. He'd stopped bucking now, he was just a mad berserk runaway over which I had no control, entirely oblivious of the dangers that lay ahead. A large timber pile under the shade of the far reaching branches of a mango tree, stood dead in our path. I yanked frantically at the side of his mouth with the bit, in a vain attempt to turn him. Then at the last split second I hacked at the point of his left shouder with the rowel of my spur and he swung angrily to the right, knocking over a heap of firewood but missing the main pile by inches. A second tree stood only yards ahead, and I flung myself forward alongside his neck just as we hit the lowest bough. It caught me high on the shoulder and completely tore the back from my shirt. Somehow I was still in the saddle and we broke cover on the far side. The worst was yet to come. A thick patch of cashew trees confronted us. There was no checking his speed, no sign of weakening, nothing could turn him from his mad devilish dash to disaster. Behind the belt of trees, not a hundred yards ahead, stretched a five-strand barbed-wire fence!

I was tempted to fling myself off, but leaving a galloping horse travelling at forty miles an hour through a bunch of trees somehow didn't appeal to me. I decided to take my chance with the wire fence!

We hit it without the slightest check in speed. The top wire caught Cang square across the front of his chest. He came to an abrupt stop and I was fired with cannon-like force over his head. It wasn't the actual fall that hurt. I came down hands first, arms akimbo, tucked my chin in and did a forward roll on my back. Forty miles an hour of impetus should then have carried me onward to my feet, and clear of anything that might be following along behind; only this time it didn't.

When Cang first hit the fence, the top wire broke under his weight, and he pitched forward on his nose. At that very moment he and I parted company. Onlookers described Cang's subsequent head stand as an impressive feat of acrobatics! No doubt I too would have been impressed, only unfortunately for me I was just about to be turned into a human pancake. Seven hundred pounds of bone and muscle came down on top of me with a devastating wallop. The air rushed out of my lungs and I felt my ribs collapse, my vision was a star-spangled sea of red.

Cang wasted no time in getting to his feet. The saddle had come adrift and dangled beneath his belly, and he stampeded forward kicking at it madly with both back feet until he shed himself of it piece by piece. Hard-riding vaqueros were already closing in from both sides. They roared with laughter as their ponies thundered by me, and shouted gleefully at each other as they swung whip and spur in hard pursuit. I had provided them with the excuse for a chase, but Cang was free now and soon outpaced them.

It was many miles before they finally put a rope on him.

V

CRACKED RIBS AND WILDLIFE

I lay face down in the grass with a tight feeling in my chest as if it were being held by a giant, invisible pair of pliers. My ribs were cracked, that I knew for certain. I'd felt them cave in when Cang landed on top of me. What a damn nuisance! I was 350 miles from the nearest doctor in Georgetown. There were only two ways of getting there. One involved a thirty-day trek on foot through the forest, and the other was by chartered aircraft. The first was obviously out of the question, and the second depended upon Jimmy being able to get through to the city on his radio transmitter, a machine which closely resembled one of the original Marconi models. He only used it about once a month and his transmissions were garbled mumblings drowned in a noise like forest fire racing through sun-scorched underbrush.

I staggered to my feet and stumbled across the demolished fence towards the house. As I walked I gradually eased myself into an upright position, breathing in short, shallow gasps. It could have been worse, I thought reassuringly. I might have broken a limb, or even my spine; but that pain in my chest, perhaps I'd punctured a lung? My shirt was in ribbons from the collision with the branch of the tree. I unbuttoned the front and let it flap in the breeze. Within a few steps it fell off my shoulders and hung in tatters from my belt.

It was an agonizing fifteen steps up the outside stairway to the veranda. Somehow I made it, resting twice on the way up, swaying against the rickety rail. I groped through to the bedroom, desperately wanting to lie down and faint. The hammock was slung as always, across the far corner of the bare room; I worked my way towards it using the wall for support. A dizziness almost overcame me as I grabbed hold of the rope, then, just in time I rolled backwards into it. As I did an excruciating pain stabbed through my chest and I screamed out loud. The curved attitude in the ham-

Longhorns on the trail

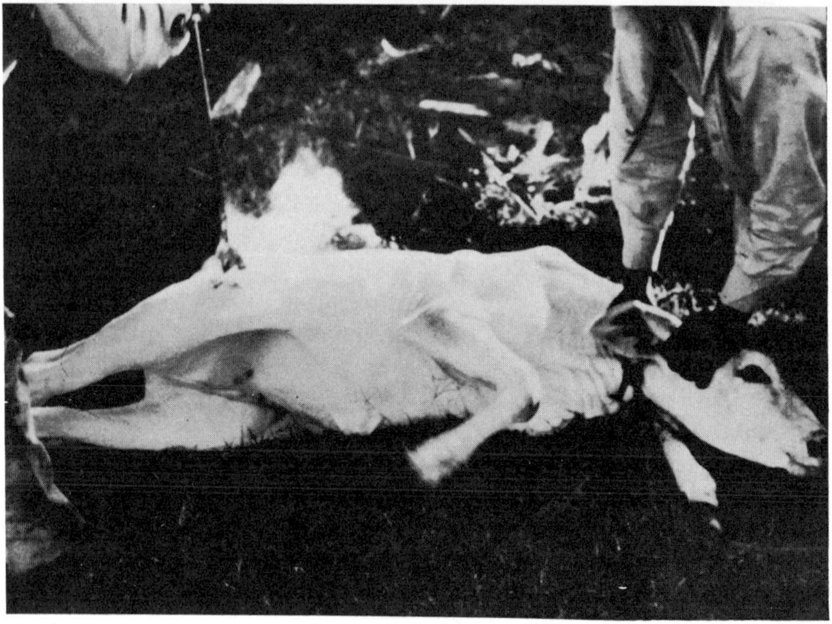

(*above*) Branding time

(*below*) Swimming the horses

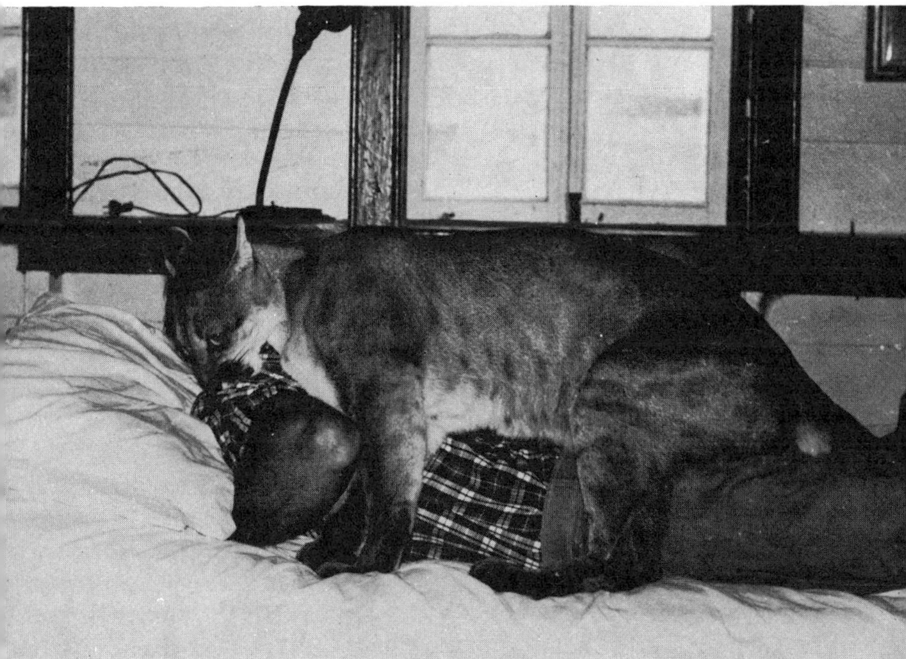

(*top*) The author with Leemo
(*bottom*) A playful bite on the shoulder
(*overleaf*) Leemo walking in the forest

(*left*) At times Leemo was strangely uninterested in horses

(*top*) Leemo was not always in a harmless mood

(*bottom*) Leemo by the Rupununi River

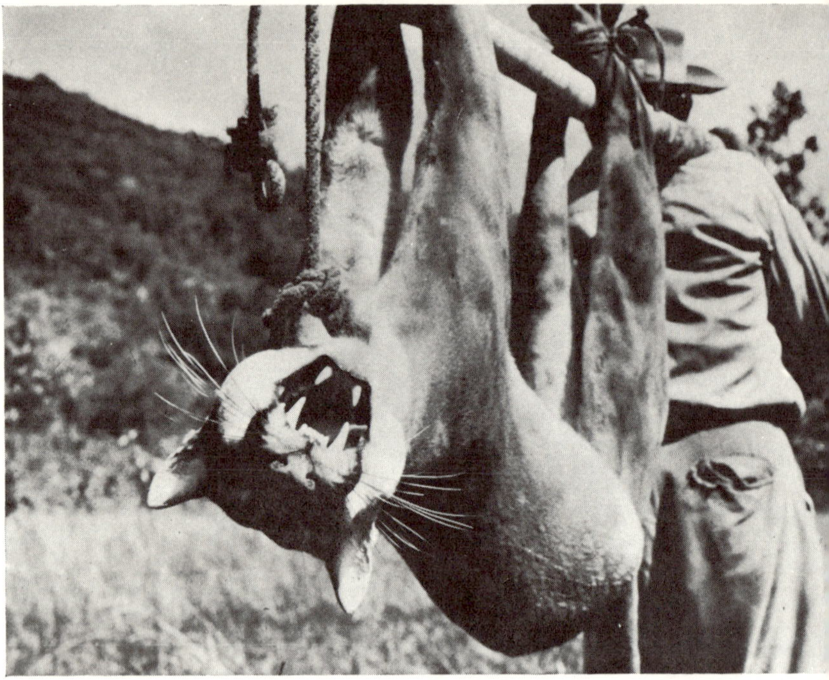

(*above*) Roland, the puma captured as a mate for Leemo

(*below*) Roland in close-up

mock had pinched my cracked ribs and further strained the torn slender muscles that held them together. Sweating with agony I wormed my way gingerly towards the edge of the hammock. My feet touched the floor first, and then my knees. Slowly I began to ease the rest of my body out.

I woke up with cool water trickling down my forehead and into my ears. Irene was kneeling beside me on the floor mopping my brow with a damp cloth. I looked up into her dark amiable eyes, and thought, 'Thank God for a friend.'

"You a'right, suh," she enquired softly.

I swallowed, almost too afraid to talk. The pain was dull now, and I was flat on my back, probably the best position under the circumstances. "I don't know," I replied hesitantly. "A horse fell on top of me and I think I've broken a few ribs."

"I heard about it," she said sympathetically.

"We'll have to strap them up—can you get some tape?" I asked.

"We got some downstairs," she answered. "I'll go and bring it."

Irene got up and hurried in her shuffling barefooted gait out of the room, returning a few minutes later with a large roll of adhesive plaster.

"We'd better do it standing up," I said. "So they'll all be in the right place."

I rolled over on my stomach and slowly rose to a kneeling position. Then, holding my breath, reeled to my feet. Stars sparkled across my vision and the room seemed to cant sideways. I felt myself falling, but Irene was there steadying me and gradually my head cleared. My ribs were swollen, but as long as I breathed carefully the pain was bearable. I felt both sides of my chest, pressing into the swelling, trying to trace where ribs should be; and if they were sticking in or out. Other than make them hurt, the examination proved nothing. Hopefully, the bones were more or less in line. The thing to do was bind them up and hope they joined the right way.

Irene applied the tape while I held onto the wall. She wrapped it around five times. It was sticky and elastic and I could feel the swellings throb under the pressure, but I felt better, as if I'd been tied back together. I tried raising my arms but shoulder height was as far as they would go. Could I walk? Cautiously, using short steps, I moved towards the door. I made it without feeling dizzy.

Then I took a deep breath, but the pain stabbed me viciously. It would be a few days before I could ride again.

In fact it was a week before I climbed rather timidly on Butterball's back. The vaqueros had released Cang back on the range. He caused too much trouble in the home paddock, fighting the working horses and breaking away from the gate whenever the boys tried to drive them into the corrals. Unless your pony is hobbled, or tethered, there is only one way to catch him, and that is by herding him together with other horses into the corral, where he can be cornered and lassoed. An animal like Cang who gives trouble to corral is a bad influence on the others; when he breaks and runs, they all break and run, and a free-running horse without a rider can run all day ahead of a dozen mounted men trying to rope him. Cang would have to wait two or three weeks until the next round-up. By that time my ribs would be healed and once more we would do battle.

In the meantime I occupied myself with quiet rides into the Kanaku mountains which stood about 3 miles from the ranch house.

The Kanakus lie in an east and west pattern and bisect the Rupununi savannahs into two fairly equal halves, Dadanawa covering the entire southern portion. The northern half is inhabited by Macushi Indians who speak an entirely different language to the Wapishanas of the south. There are also a number of small ranches owned by settlers who took Indian and Brazilian wives. To the west flows the Tacatu River, a northern headwater of the Amazon, and also the boundary between Guyana and Brazil. A gap some 5 miles wide between the river and the mountains provides a gateway between north and south savannahs. On the eastern flank almost trackless forest stretches without a break to within a few miles of the Atlantic seaboard.

The Kanakus are covered in dense jungle which grows miraculously on granite boulder-strewn slopes, rising to 4,000 feet. Through their midst the black waters of the Rupununi river meanders lazily 100 miles to join the Essequibo.

Although this fortress defied penetration on horseback, the persistent savannah wormed its way between the foothills in countless valleys and inlets. Each of these Butterball and I carefully explored. I talked to him about the birds, the animals and the trees. He listened patiently and wisely, and when he'd had enough he would turn around and head for home. Invariably I applied the prescribed aids of bit and heels to dissuade him; but Butterball's determination always won when sore ribs dictated that I give in.

The bird life, particularly along the wooded river banks, constantly amazed me. Tiniest and most beautiful were the humming birds, hovering, ascending, descending, flying forward and reverse like miniature helicopters, probing with stiletto bills the nectar and insects from a banquet of tropical flowers. They ranged from the minute bronze Coquettes, little more than two inches in total length, to the five-inch purple-faced Jacobins, with their metallic green crowns and white tails. Occasionally, I came upon their nests, thimble-sized delicate structures of plant fibre wedged in the fork of a twig. They were often ingeniously tethered in place with a tangle of white thread which, on closer inspection, proved to be stolen cobweb. Two dainty white eggs were the usual clutch.

White-throated toucans were common. Poor flyers, they flew shakily with erratic wing beats from tree to tree as I rode by. Their grotesque bills were as long as their bodies, and they called at me with musical bell-like tones which piped up and down the scale.

Birds of prey were everywhere. One of the larger species was the savannah hawk, rufous red in plumage, standing about twenty-five inches tall on yellow legs. They spent more time perched in trees, watching like sentries, than they did flying. One day I chanced upon an exciting duel between a savannah hawk and a mussurana. The mussurana is a non-venomous snake that makes its living by eating other snakes. It is steely blue-black in colour with large scales, and grows to lengths of seven feet. The mussurana is strong and agile, more than capable of throwing a muscular coil about an impudent and incautious hawk.

I was attracted to the scene by what appeared to be a large injured bird floundering in the long grass. However, on drawing closer I spotted the black mussurana with head poised at least 2 feet erect in a striking attitude, confronted by the sparring hawk. The two faced each other about 30 inches apart; the snake waved its head menacingly from side to side while the bird bounced up and down like a ball, wildly flapping its wings. From 40 yards I had an excellent view of the proceedings and the hawk took not the slightest notice of my presence. His strategy seemed to be to draw the mussurana's fire by feinting with an outstretched wing, leading with it towards his opponent as a boxer would do with a fist. This action caused the snake to strike at the wing. As it did so the hawk bounced 4 feet in the air, fluttered over the snake's head and lashed out with his talons as he passed and landed behind it. Immediately the mussurana recoiled and struck again, but once more the savannah hawk deftly avoided the stabbing head. This was the deadliest game of leap frog I'd ever seen. Should the snake have got a fang-hold anywhere on the bird it would have wrapped him like a python in its coils and slowly squeezed him to death. On the other hand, the reptile's head was being bombarded by the hawk's needle-sharp, lethal talons. For ten minutes the duel continued, the combatants see-sawing back and forth jockeying for position. Finally, the burning sun and heat from the white sandy soil began to take its toll of the mussurana. Gradually, the strikes became slower and weaker until its head would fall to the ground and wait two or three seconds before recoiling. But the savannah hawk seemed as fresh then as when he first began. Grimly he slashed at the faltering black head. He knew the end was near, to hold now with his talons would be foolish, the snake could still wrap him with its last ounce of strength. Mercilessly the hit and run tactics continued until their macabre end. A few minutes later the vanquished mussurana lay dead in the sand and the triumphant hawk busily indulged in his hard-earned meal.

During the ten days of regular rides along the foothills I saw few signs of larger game, other than tracks of white-tail deer. Occasionally, I caught a fleeting glimpse of a white rump as one bounded away with a snort through the trees. One morning when the sun was little more than half an hour old and dew still clung to the tall grass, I rode Butterball to the crest of a hill which had been a frequent observation point. About a mile away, twenty or thirty vultures were circling in a spiral. Gradually, one by one they peeled away and landed amidst a belt of trees only a few hundred yards from the mountain side. Cattle and horses frequently ventured into the valleys, for the rich soil, washed down from the mountains in the rainy season, produced better grazing than on the open plains. Realizing that one might be dead, or perhaps seriously injured, or sick, I spurred Butterball and cantered down the hill to investigate.

As I drew closer I spotted two or three dozen black vultures perched about several trees. This sign invariably meant that either the prey was not yet dead, or that some other animal was present at the carcass. I slowed Butterball to a walk and approached cautiously. To my astonishment, as I skirted some intervening low brush, I came upon their quarry only a short stone's throw away. It was a freshly dead horse and standing on top of it were two enormous black and white birds. These I recognized immediately as king vultures which, save for the condor, are the largest birds of carrion in the Americas. They were a fantastic combination of the beautiful and the grotesque. Each bird stood almost 3 feet tall, with sparkling, snowy-white shoulder and breast plumage. A black trailing edge to the wing, and black tails added a striking contrast. Their heads were bare, revealing purplish skin; the neck orange, surrounded at its base by a slate-grey ruff of feathers. A large wattle of orange flesh protruded from the base of the beak. To complete this bizarre handiwork of nature, black beady eyes were ringed like the bull in a target with white and red. My observation was terminated within seconds for the birds spotted me and flew off without delay. Their powerful wings spanned at least 6 feet and they were soon high up over the mountains. In contrast to their shyness, the black vultures descended at once on the carcass and began tearing at it furiously. In a moment the dead horse was a seething mass of fierce, hissing, heaving, struggling feathers. Obviously, despite their overwhelming superiority in numbers, the smaller black vultures were afraid of the kings, and for that reason would not go near the carcass while the latter were feeding.

I trotted Butterball towards them and they flew off to their nearby perches with a deafening rattle of wings. My prime purpose was to determine why the unfortunate animal had died.

Being aware that jaguar and mountain lion were common in the area, I naturally suspected that one of these cats might well have been the cause of its death. My suspicions were soon confirmed. Just above the horse's eyes were two punctures. Behind the ears on the upper part of the neck, two more. In addition, there were extensive claw marks down the animal's side near the flank. This was a classic example of the jaguar's work. Although it was the first one I had seen, the Indians had told me that the spotted cat always killed by biting its prey through the top of the skull, causing direct injury to the brain. On the other hand, its rather smaller cousin, the mountain lion, kills by gripping its quarry's throat, death resulting from asphyxiation. Hence, the labour of each can always be distinguished.

There were no signs of a struggle and the squabbling vultures had done a clean job of sweeping the surrounding sand with their wings, obliterating any spoor. Nevertheless, a path of beaten-down grass revealed that the horse had been dragged for several yards into the trees. When I followed this path I found clear signs of where the kill had actually taken place. Over an area of 15 square yards all grass and shrubs had been flattened. The horse's hooves had gouged deep furrows into the ground as it fought in vain to shake off its attacker. Yet, the battlefield was sparse for an animal the size of a jaguar matched against a 700-pound horse. Here was an illustration of sledgehammer power, as the assailant had obviously dispatched its prey with uncanny ease. Searching further I soon found cat tracks in the sand between the grass roots. I tensed with excitement, and dismounted to inspect them more closely.

They were clearly defined in the damp morning soil. Four toe prints and an oval-shaped heel.

I led Butterball back to the dead horse and began to comb the ground for further spoor in a radius of a dozen yards about the kill. In five minutes I had picked up the trail and began to follow it towards the thick forest that began at the foot of the hills. It was not difficult to do, the grass grew in scanty bunches and the low morning sun threw a shadow across the indentations of the pug marks, making them stand out. My heart pounded as I moved forward into thicker and thicker cover with the rash enthusiasm of a successful beginner.

Butterball followed me obediently a yard behind on the halter rope. He tugged occasionally at a choice morsel of grass as we went by, and seemed totally unconcerned that one of his kind had been struck down only an hour or two before by a deadly marauder whose smell still lingered in the air.

Just ahead a steep-sided gully blocked our path. It was about 6 feet wide and as many deep. Pronounced splayed out spoor clearly indicated where the jaguar had leaped into it and continued his way up the bed towards the mountain. I followed him for fifty more yards to a small pool trapped between a cluster of smooth rocks.

Here the cat had drunk, splashing water about on the boulders, for they were still wet. The pool was muddy, and I began to wonder just how long stagnant water might stay muddy after it was stirred up, and how long before splashes would dry on the surface of a rock—half an hour, an hour, three hours? Around the pool, the banks of the gully had been broken down by cattle and horses. Here the cat had climbed out and the trail pointed through knee-high grass towards the forest. I pulled Butterball impatiently across the ditch. There was no need to look for pug marks now. Before me the expanse of tall feed, combed by a gentle breeze, took on the tranquil look of a gently rippled lake. But the surface had been disturbed by the transient predator, and a tell-tale rivulet of broken grass marked his path.

I followed swiftly in the wake, jerking intolerantly at the halter rope, compelling my reluctant pony to break into a shambling walk. The tough-rooted iron grass got taller and taller as we neared the forest edge. All at once it was waist high, then suddenly I was wrenched to a halt.

Butterball was anchored to his tracks, leaning back on the halter. Both ears quivered forward and wide open eyes looked right past me, focused on an unseen object that lay ahead.

I looked in the direction that he was staring but could see nothing because of the high grass. Immediately a sense of foreboding gripped me. What was I doing tracking a dangerous animal into thick cover, unarmed, and worst of all, alone? To this point it had been an exercise for I was keen to learn the ways of the hunter. The thought now flashed across my mind that I was trying to learn too much too quickly.

Slowly, I stepped two paces backward alongside Butterball's shoulder. Without turning I lifted my left foot, crossed it over my body and groped for the stirrup. Finding it, I clutched it firmly between my bare toes, slid my right hand up to the pommel and gently levered myself into the saddle.

Even before I was squarely seated I saw the cat. He was facing us about twenty paces off, a great glistening gold and black animal with shoulders on him the width of two bulldogs. I had seen jaguars in a zoo, but was completely taken aback by the savage, gleaming splendour of this beast. His mouth was slightly open, just enough so that I could see a few inches of pink tongue and the awesome yellow-stained canines that a few hours before had crushed the skull of a full-grown horse. I had expected a lithe, graceful animal, but here was a bulky, craggy-chested creature with huge forearms, broad jowly head, and a hump of muscle on his withers like a range bull. He shifted uneasily, moving his head from side to side as if aware that other opponents might be closing in on him from the flanks. His tail swished gently and the black tip twitched independently and convulsively. Yet I could see that there was no intention in his eyes to attack, only majesty, pride and power. Slowly he turned sideways so that I could see the magnificence of his rosette-spotted hide, and then melted away into the long grass.

The incident had an indelible effect on my whole life from that day forward. Until then I had been completely absorbed in the problems of becoming a competent vaquero and striving for recognition as such amongst the Wapishanas. These tasks were so formidable that I had almost forgotten my boyhood passion for wild animals. The gurgling streams and artful fox of my schooldays had been drowned under the beckoning of the thundering herd. Now, suddenly again, my feelings for nature were revived. I saw my life at Dadanawa as an unparalleled opportunity to study the fauna of northern South America at first hand. It could be done almost without trying. I had 3,000 square miles of wild natural game reserve at my disposal and the most ideal of transportation—the quiet, unobtrusive horse.

These months saw not only the moulding of a vaquero, but also the beginning of myself as a naturalist.

(*above*) Climbing Shiwishi-Win with Mr Manager

(*below*) A cattle killer brought to bay

(*above*) Mr Manager and family. (*right*) A jaguar

(*above*) Shadow

(*below*) A snack at the store-room for Choo-Choo and the Margay

Choo-Choo would never allow the Margay to stray too far from her side

(*above*) Choo-Choo

(*right*) Choo-Choo at the store-room door

(*top*) Nothing of a chewable nature was safe from Choo-Choo

(*bottom*) There were times when Chico could be dangerous

VI

ROUND-UP

It was November and my ribs had healed perfectly. The wet season finally was at an end. Fifty inches of rain had fallen since the beginning of the year. Night and day we had been bombarded by clouds of biting flies. From dawn to dusk the kabouras swarmed into eyes, ears and nostrils like wind-blown sand. At sundown their dastardly chore was taken over by mosquitos, flying hardback beetles, cockroaches, sandflies and flying ants. The light from the smallest hurricane lamp drew them like a magnetic field. To write or read was impossible, there was no rest or escape, meals were eaten in the dark, and tired vaqueros flopped miserably into hammocks, to lie tormented in the damp, still, oppressive heat.

But now the plague had passed. Blue skies yawned overhead and turbulent rivers had settled once more within their banks. The savannahs, drowned for five months beneath the floodwaters, sprang alive in a verdant carpet of luscious grass. Thirty thousand Longhorns marched hungrily towards the valleys from their haven of stony hills. Now lakes and waterholes were full, and gaunt and mangy cattle would graze unimpeded for seven insect-free months.

This was round-up time. I hadn't seen Jimmy Brown for weeks. There were more than a dozen corrals scattered at strategic outposts on the ranch, and he was busy making them ready for the big herds with last-minute repairs.

Round-up sums up the economy of an open-range cattle ranch almost in one word. It is the cardinal event around which the whole operation revolves. Thirty thousand wild animals were out there, scattered across 3,000 square miles of rocky hills, drained by a multiplicity of streams and swamps, and littered with thousands of islands of impenetrable brush. Every head had to be hunted for, chased, many of them lassoed, and finally driven as a herd into the nearest corral. As many as 1,500 Longhorns might be packed into the corrals in one day. They would be driven through a chute and separated into smaller yards. Young animals would be branded, ear marked for identification, and castrated. Mature steers would be weeded out and isolated. These would later be driven 350 miles to market in Georgetown. Each animal would have spent from six to eight years on the Rupununi Savannahs developing to its full size and strength. The mountainous jungle trail from Dadanawa to the coast would test every steer's stamina to the extreme. Many would not survive the journey but the revenue from those that did was the only reason why the ranch existed.

Daily, Indians filtered into the Dadanawa headquarters looking for work. Charlie decided which ones to employ, and I wrote their names down in a notebook. As the only one who could read and write I would be responsible for paying them, and forwarding the information to the Georgetown office on one of the charter flights.

"This man throw lasso good," said Charlie. "Put his name down for round-up."

"This man worse than useless," he commented graphically about another. "Put him down as cook."

There were three columns of names. The first numbered sixteen, including Charlie and myself. We were the vaqueros who would actually go out and catch the cattle. The second column numbered twelve. These were the 'pastoriadors', made up mostly of men who were too old to stand up to the hard work of the round-up, and youngsters who were just learning. Their task was to graze, or *pastoria* the steers caught on the daily round-ups. From this herd, animals would be selected for market and driven to Georgetown. The ten men in the third column were responsible for that job, and called 'boiaderos', derived from the Brazilian word *boiado*, meaning beef herd.

The boiaderos drove the cattle on horseback north for nine days across open savannah to the forest edge. At that point they abandoned their ponies and accompanied the herd on foot along a narrow trail cut through the jungle. This phase took sixteen days and involved swimming numerous rivers. An added hazard were stampedes caused by prowling predators, with cattle quickly lost in the dense forest. The trail ended at the Berbice River where the steers were driven aboard a steamer and transported the remaining distance to the coast. The men then returned to the savannahs on foot, picked up fresh horses and a new herd. Boiaderos were scorned by the vaqueros who considered themselves the elite, but that was their opinion and not a fact.

By the end of a week we had a full complement of men. Many had come with their wives, travelling barefooted as far as 50 miles from some of the outlying villages. Each Indian was given an advance against his pay in trade goods from the ranch store, and the women returned alone to their homes carrying supplies of cloth, knives, axes and iron cooking pots. Then every man was issued his equipment: a saddle, horse blanket, bit and one raw cowhide. From the latter he was expected to make his lariat, halter rope, bridle, reins and hobbles.

Cowhide is of such a nature that it can only be worked at night, when it is soft and supple from the damp air. In daylight it quickly becomes as hard as a piece of wood. An obvious solution might be to wet it in water, but such treatment never creates uniform pliability and it becomes impossible to cut and plait tightly and neatly. Two nights of ceaseless labour from dusk to dawn was sufficient to plait a 100-foot-long lasso. A third night took care of the halter rope, hobbles and reins. By this time even the toughest Indian hands were raw and bleeding, for it required strong pulling on each strand of the trimmed hide to lay the braids neatly together.

Fortunately I had learned the basics of the art during the preceeding months, and by working for two or three hours a night over a period of weeks, provided myself with strong new equipment with a minimum of pain. Charlie was a great lasso maker and all his gear was beautifully made with the stamp of a real professional. He worked every night in the barracao under the dim light of a hurricane lantern. I watched him often and made a point of freely commenting on the excellence of his craftsmanship. At the same time I acquired a cowhide for myself and began cutting it up into strips. My clumsy efforts must have so appalled the foreman that in a rare moment of rapport he left his own work and with great care demonstrated the correct way it should be done. From then on, he kept a critical eye on my workmanship in his own inimitable way. There was never a nod of approval if I was doing things correctly, but always a reproachful grunt when I went wrong.

A lasso is made in five stages. First, the green hide is stretched by pegging it out on the ground in the sun. After two days it is dry. Then wood ash is scattered over the hair. This acts as an abrasive and the hair can then be scraped off using the sharpened point of a stick. The strongest part of the skin is in the hind quarters. These are cut out into two oval shapes of equal size. Then the worker sits on the ground, and, starting at the edge of the oval, cuts round and round in a circle, producing a continuous strip about a quarter inch in width and approximately 200 feet in length. The second oval is treated in a similar manner. These two lengths halved in the middle produce four strips, each 100 feet long. In the fourth stage, the sharp edges of the strands are bevelled using the point of an extremely sharp pen-knife. This causes them to lie snugly together during the final stage when they are plaited. The braiding method is similar to that used on a girl's hair. While working, the lasso is tied to a post, and the vaquero, standing up, passes each strand in turn around his waist and throws the weight of his body against it to draw it tight.

The next step in the round-up process was to collect all the horses. This took the better part of one day, for although they grazed nearby they were scattered across a 50-square-mile valley between the Rupununi River and the Kanakus. We rode out at 3 a.m. and reached the far end of the valley at sun-up. There, we split up into five groups of two and gathered the horses, about 150 in all, on our way back to the corral. Later that afternoon we assembled, and Charlie allotted our mounts.

Each vaquero was given four horses. The work on the round-up was so arduous, and the distances to be covered so great, that a horse could only be ridden effectively for one day, followed by three days of rest. The pastoriadors and the boiaderos had only one horse apiece. They were expected to herd the cattle and drive them at a slow walk, and therefore one pony was considered adequate. Their mounts were invariably elderly and no longer usable on the round-up. Unfortunately the herds did not always agree to be moved at a slow walk and stampedes were common. As a result there were many dissatisfied comments from the herders as their horses were allocated.

My string included Butterball and Cang, together with two bays, a colour which I had now learned to call *castano*. One was a tall slender animal with good chest, but unusually slim flanks. His teeth showed him to be about 4 years old. I called him Guitar because of his pinched-in waist. He was quiet, with one year of cattle experience behind him. The fourth was a big, nervous, mule-shaped colt, with a slender tail, sharp protruding breast bone and white walled eyes. He turned out to be incredibly strong and fast, and also the only horse I ever rode that was frightened of his own tail. As it swished sideways, he would see it out of the corner of his eye and leap away from it, simultaneously drawing it between his legs like a cowed dog. Despite this he was full of courage, and I called him Matador.

There were forty-odd Indians gathered in front of the bunk-house as I prepared for my second ride on Cang. The bubbling laughter of their conversation, coupled with descriptive gestures, made it blatantly obvious that they watched in fervent hope of seeing my neck make speedy contact with *terra firma*. My stomach was knotted with fear and I fidgeted nervously and unnecessarily with my equipment. The first time I rode Cang I had been frightened, but now I was terrified. I began to regret that I had ever told Charlie I wanted to ride him. There is an unwritten law amongst the vaqueros that if a man once agrees to break a wild horse and puts his saddle on him, he must see it through to the bitter end. To refuse to ride him again after being thrown would be tantamount to losing one's nerve. The worst disgrace, and a man would wear it for the rest of his life, was to concede defeat and allow another vaquero to break his horse. It is alright to refuse to ride a particular animal in the first place, at this people laugh but quickly forget. But to ride once, and then refuse to ride again, earned the title *kit-chen-ait*—a coward.

To the Indians I was an outsider who couldn't even ride Butter-ball without falling off. They had never seen a white man who could make a vaquero, and my increasing knowledge of the Wapishana language fed me information that made my cheeks burn. They called me *Meu-tirryan*—the useless one!

Yet no Wapishana had ever ridden Cang, he was the best chance I had to prove myself.

There was the usual problem in getting the saddle on, and even worse difficulty in trying to climb aboard. In spite of the blindfold Cang knew at once when I tried to put my foot in the stirrup and lunged forward, kicking out viciously with his hind legs. Kicked to death by a horse is for the cowboy one of life's less pleasing thoughts, and required preventative action. I discovered that by holding the nearside shank of the bit in my left hand and pulling it closely into his shoulder, he was forced to move in a tight circle around me. All I had to do was stand in one spot and turn with him. Unfortunately there were two drawbacks. One, he had a tendency to stamp on my bare feet with his front hooves; and two, he circled me at a mad gallop. The latter not only made me dizzy, but it was also incredibly difficult to turn round and round at high speed hopping on one leg trying to get the other foot in the stirrup. Still, somehow I managed, and swung into the saddle.

Before I could rip off the blindfold Cang plunged his head between his front legs and took off in a frenzied run. I yanked and jerked on the reins for all I was worth, but to no avail. Fortunately he ran in a straight line through the various gaps between the buildings and out into open country. I gave up trying to stop him and let him go. After a quarter mile he raised his head and I jerked off the blindfold. He ran for a good mile before he began to wear down, and as he did I brought him under control. By this time he was blowing hard and dripping in sweat, but I realized that now I had the upper hand.

I took him on a three-hour ride across the savannahs and then turned him for home. He had a head like iron which he carried low between his front feet. It yawed from side to side and I was at constant battle with the reins until my hands were raw and bleeding. He seemed to shy at every object that lay in his path, armadillo holes, ant-heaps, even a harmless stick. His ears would go forward, air would roar through his nostrils, and then with a giant leap he would hurl himself sideways, shooting a flying kick at the object for good measure. But he was beginning to respond. I could turn him to right and left and stop and start at will. His canter was clumsy, but despite his head being between his front hooves, he never once showed a tendency to trip. At a gallop he could run like the wind.

About 2 miles away from the ranch house I spotted three riders approaching. As they drew closer I recognized one as Charlie, accompanied by two of the men. I felt uncomfortable as I rode up to them, but Charlie I noticed, wore the faint trace of a smile.

"We were looking for you," he said, with genuine concern. "How's your horse?"

"He's fine," I replied, trying to be nonchalant.

Charlie nodded his head in quiet approval. Suddenly there was a loud thwack and Cang rocketed forward with an angry squeal in a wicked, twisting buck. I hung on grimly and kept my spurs clear of his flanks. In four jumps I brought him to a stop and turned. The Indians sat back on their ponies and roared with laughter while one mischievously waved at me with his rawhide whip.

"*Vaquero airee* [He's a vaquero now]," said Charlie affirmatively.

"*Meesh vaquero* [It is true]," agreed the others with a friendly smile.

They moved their ponies alongside mine and together we rode off to Dadanawa. Even Cang seemed proud; at last the devil was finally on the road to becoming a vaquero's horse.

During the following months our horses scoured every mile of the Dadanawa range in pursuit of the widely scattered herds. With our hammocks on the back of our saddles we moved from outpost to outpost. At night we camped by quiet streams under the stars, and by day we spread out in parties of four and hunted the wily Longhorns through swamp, brush and hills.

The whole operation was a living museum of practices that died out in the western United States at the turn of the century. Except during round-up time, the cattle had virtually no attention and were left completely to their own devices. Bulls fought and killed each other for cows. When they were either too old or too young to win they joined together in groups, and bands of twenty and thirty were common, roaming the range divorced from the breeding herds. If 3,000 head were to die of disease in a single year the loss could conceivably go by unnoticed. There were at least 5,000 animals on the ranch that were so wild they were impossible to catch. The herd of 30,000 was the number that one could normally expect to corral and count. Yet the only remunerative way to run the Dadanawa was on this shoestring basis. The soils were poor, heavy annual rains bleached them to a white sand devoid of all essential minerals. A carrying capacity of only ten head to the square mile made intensive concentrations of closely supervised herds impossible. The only communication with Georgetown for the purpose of bringing in supplies was by air. Freight rates were high, making all heavy items such as fertilizer and supplementary mineral feeds completely uneconomical. This made virtually any type of improvement extremely expensive. The final cold analysis was whether the price of a slaughtered beef was more than what it cost to grow. Until such time as a road or railway connected the ranch with the Georgetown market, Dadanawa would have to stick with her primitive, but profitable, nineteenth-century methods.

Every round-up had a 'rodeodor', a pre-assigned spot to where the running animals would be guided, and gathered into a controllable herd. The rodeodor was usually in a stretch of open country. It was like the hub of a wheel with spokes of cattle thundering into it from all directions.

From that point the herd was driven by twelve men, with two fanning out on either side looking for more animals. Where possible, they rode the high ground so they could scan as much country as possible. Both the men driving the main herd and those covering the flanks left a constant string of burning grass behind them. As a result, though they might be riding across low ground, they always knew each other's approximate whereabouts by the smoke on the horizon.

A single round-up usually covered a stretch of country 20 miles in length and 3 miles wide. Depending on the wildness of the cattle it could take anything from six to twelve hours to complete, from starting point to corral. The size of the herd also varied considerably. Poor arid areas might only produce 300 animals, fertile ones over 1,000.

Once inside the corrals the herd was driven through a parting chute. Steers for the trail were separated into one pen, those for branding into another. Breeding cows, bulls and immature steers were also sorted. The toughest job was then the actual branding. Every animal had to be lassoed and thrown. The vaqueros worked in pairs, one catching the head, the other the heels. When the catch was stretched out on the ground a third man drew an iron from the nearby fire, and burned the Dadanawa brand, a double H on its right shoulder. A fourth vaquero cut a V-shaped notch from the right ear. This notch provided ready identification of a branded animal on the range, for it could be more readily seen than the letters on the shoulder. Bull calves were also castrated immediately. The entire operation took little more than ten seconds from the time the ropers threw it to the ground.

The unbranded animals were called 'orrelludes', from the Brazilian meaning 'long ears', or un-notched. A little more than half were in their first year, but there were also many in their second, and some that were several years old. The latter were wild, and known as 'lavraderos'. They had avoided capture on all previous round-ups. Cornered, the lavraderos were vicious, mean and dangerous, often armed with rapier-sharp horns. Extreme caution was required while working in the corral. They charged us constantly, there were narrow escapes daily, and injuries commonplace.

In some parts of the ranch the lavraderos were such a problem that they could only be captured on special sorties. A wild Longhorn is one of the most intelligent of animals. Sight, hearing and sense of smell are all acutely developed, and keen over phenomenal distances. During the regular round-up the bellowing herd on the move would warn the lavraderos 3 miles ahead that vaqueros were approaching. The outlaws then did one of two things. In open country they began to run, and, with 3 miles start, usually managed to keep ahead of the vaqueros. On the other hand, if there was a thick patch of forest nearby they entered this and hid, often until nightfall.

Lavraderos that ran for cover were called 'matteros'. Matteros were the most cunning, tactical, manœuvring animals on the Dadanawa Ranch. They could only be caught by the swiftest horses and lassos. Many died of old age without a rope or brand ever touching their hide.

Special expeditions were planned to thin them out, usually before or after the regular branding season. Finally, after four months of daily round-ups, when horses and men were honed down to the edge of their best performance, Charlie announced that the time had come to catch matteros.

It was March. The blistering sun beat down upon the Rupununi savannahs as it had for the last five months without a drop of rain. Cattle and horses collected around rapidly disappearing water holes. Some bogged down in the soft mud and were trodden ruthlessly into the quagmire by others who struggled towards the last drops of evaporating moisture. Vultures stood on the heads of the exhausted living, plucking out their eyes, while waiting for the rotting hides of the dead to soften for their beaks. If they were fortunate, a vaquero might happen by, and with the help of horse and lasso, extricate some of those not too hopelessly set in the bog. But the range was vast, and waterholes too few and far between to be visited regularly. The Rupununi is littered with the bones of tens of thousands of cattle and horses who finally lost their lifelong battle with the elements.

With water scarce, and cattle grazing close to it, we planned an attack on a particularly crafty band of matteros that inhabited a thickly forested region close to the headquarters.

I took up the last inch of slack in the girth, and groped for the mug of black coffee that the packman handed me. The morning star just peeped above the inky eastern horizon. Three o'clock and barely three hours ride to where the matteros were living; we would make it by sun-up.

A shape loomed alongside me. Charlie, already in the saddle. I threw away the remnants of coffee and unhitched Matador from a nearby tree. The silence was broken only by the chinking of bridles and creak of saddle leather as sixteen men rode quietly away into the darkness.

There was a cold nip in the air and I soon buttoned my shirt collar at the neck. The coffee glowed pleasantly in my empty stomach; we never ate before riding out at an hour like this and the hot liquid felt like a blood transfusion. I kept my pony just behind Charlie's, for there were many narrow trails through dense trees and one had to be careful of low overhanging limbs. After many hours of continued night riding a horseman appears to develop a sixth sense about unseen objects. He seems to feel that a branch is looming ahead of him and either ducks or throws up a protecting arm in front of his eyes. The horse is blessed with a similar faculty but, unfortunately, only applies it to his own body, and trots happily under the boughs of trees that cut his rider off at the middle.

Today I felt particularly proud to be riding with the Indians. Charlie had only selected the best vaqueros to accompany him, and the others had been left at the ranch to help the pastoriadors control the main herd. It amounted to complete acceptance, and I now knew that the mistrust which had so plagued my early relationship with the Wapishanas had finally gone. All of us were *camarades* together.

We arrived half an hour before it was light enough to make out detail at any distance. Our approach had been carefully planned downwind and we dismounted about a mile from where we expected to find the matteros. Charlie and I left our horses with the boys and went ahead on foot to do some scouting. By the time we had climbed a high rocky outcrop that gave good all-round observation it was light enough to see.

Sure enough, about half a mile away, quietly grazing amongst a belt of trees, was a bunch of wild cattle. There appeared to be about fifty of them, including many large steers.

Suddenly Charlie hissed and pointed excitedly.

"*Garotch di'kurd* [a particularly dangerous bull]," he whispered, nudging me violently with his elbow. He went on to explain that he recognized one of the animals in the herd as a particularly notorious renegade bull. The animal was grazing apart from the others. It was an enormous black-bodied brute, with white face and wide-sweeping horns. Like all range bulls the tips had probably been honed on boulders to needle-sharp points.

"Two year ago vaquero lasso him," Charlie breathed. "He break lasso and kill horse. Maybe today we catch."

With that Charlie crawled back over the skyline and we both hurriedly returned to the horses.

We rode at a slow walk, single file, hugging close to the forest edge. The cattle were grazing among thinly scattered trees about 400 yards from the nearest dense cover. The idea was to cut off their inevitable dash towards the forest and lasso as many as we could before they got there. Nothing on earth would make them run in the direction of the open savannah. Even a solid wall of horsemen would not turn them. Every one, unbranded, untamed, was going to come our way, and only the toughest rawhide and fastest horses were going to stop them.

I untied the rope from the back of my saddle and passed it into my left hand. Quickly I reached down and checked that both girths were cinched up tight. I gave the front one an extra tug and made sure that the end of the lasso was securely anchored to the ring under my right thigh. All the strain would be taken at that point, and if it was slack I could have the saddle jerked off the horse under me.

We were still a quarter of a mile from the cattle when without warning an old cow standing 'sentry' spotted us. She started for the bush at once, and the rest stampeded after her.

Sixteen horses surged forward in a broken line. Matador lunged quickly into the lead alongside Charlie, while a few of the other fast ponies ran neck and neck on either side of us.

The white-faced bull headed the bunch with four big steers and the sentry cow. A mixed quantity of smaller bulls, steers and cows followed twenty yards behind. Our horses quickly overtook some of the stragglers, but they were thin old cows, not worth stopping for. Every vaquero had his eye on the leaders.

Slowly we gained on them, but they showed little sign of tiring. Now only 300 yards separated the matteros from their forest refuge.

I had dropped behind a stocky grey pony named Bandero. He was a fine, swift-footed animal, one of the best on the ranch. Suddenly he tripped and fell awkwardly at full gallop. He was underneath me before I could swerve. I heard his neck crack like a pistol shot, and Matador bounded over his flaying hooves, missing the thrown rider by inches. Looking back I saw the Indian already on his feet and running for a tree, for there were cattle following, and in poor humour.

Three of us were drawing up on the leaders, while the other vaqueros had already started to lasso some of the bigger stragglers. Their horses were blown and they knew they had little hope of catching the stronger animals ahead.

I was level with the white-faced bull. Charlie was swinging his rope at a black, rangy, ring-eyed steer, and our companion peeled away after another. The bull was mine, and only 150 yards left in which to catch him.

Keeping my eye on the quarry I shook out a large loop in my right hand, for I would need a long throw. It was fortunate I did, for without warning the bull broke left and came after me, meaning business. Just in time I sank a spur hard in Matador's flank. His extra burst of speed put much needed inches between his rump and a wicked horn which stabbed harmlessly in the air as the bull flashed by.

Cautious now, I hung well out to the brute's right, steadied my pony down, and let the mattero draw slightly ahead. We passed through a group of trees, and there was an open spot. I gave Matador both spurs and swept in at a sharp angle to the Longhorn's rear with all the speed I could muster. A quick swing of the rope as I closed in, and then the throw as I passed within 10 feet of his tail. The loop, like an extension of the arm, shot out, encircled the horns and snapped shut. The fish was hooked.

As the rope hit, the bull swung about and came after me. This time he wouldn't let me escape so easily. Matador was going flat out and I was using whip and spur. His life and mine depended on me getting the most out of him. We were heading for the forest and 50 yards ahead lay a thick belt of trees.

I had no choice but to sacrifice speed and make a sharp left turn. The bull's horn hooked Matador's tail and he scorched on by, straight for the woods. Rawhide was screaming through my fingers as we neared the end of the rope. With only a few feet left I dragged the horse's foaming mouth hard to the right and began to haul him to a stop. I had to throw the bull hard and knock out his wind. Then hope he'd stay down long enough for me to dismount and tie him up. Three coils of lariat remained in my hand; I tossed them clear and lurched to a halt.

Fearing the rope would snap and blind me with its devastating elastic recoil I threw up an arm and shielded my face. There was a tremendous jerk, Matador groaned and gave 5 yards under the strain. At the same time the bull hit the ground like a wet sand-bag dropping from a four-storey window. But the rawhide held. I dismounted immediately, and leaving Matador to keep tension on the rope, raced forward.

The mattero was lying on his side facing my horse. Circling him slightly I approached from the rear. My objective was a quick tie around his back legs before he could recover from the fall. As I put a hand on him he reacted with a lightening backward kick. It was so unexpected and fast that I was caught completely unawares. The hoof struck me a terrific blow in the solar plexus and threw me spread-eagled on my back. In an instant the wild bull was on his feet, and coming straight for me.

I barely rolled aside in time as his horn dug a deep furrow in the ground where I had been lying. He would have got me with the second thrust had Matador not stood his ground and held him back on the rope. He was only a few paces from me, straining at the lasso, eyes blazing, tearing up the ground with his front feet. His head tossed in the air, sweeping his horns viciously at a target just beyond reach. With every lunge his nostrils threw solid clods of foam high in the air.

I couldn't get back to my horse. That meant a foot race between the bull and I, and me an odds on loser. The only thing to do was keep his attention, in case he should decide to attack the horse. Improbable as it may seem, cow ponies have a very bad habit of standing still in the face of a charge, when they haven't got a rider to tell them to move.

Suddenly I heard the sound of approaching hooves. It was Charlie, galloping up behind me rope in hand. But an immediate solution was far from in the offing. If Charlie got too close and swung his lariat, the movement might frighten the bull and cause him to turn and run, putting Matador in great danger. The only thing to do was have Charlie occupy the animal; while I backed further away and attempted a wide detour back to the horse. The foreman agreed. He walked his pony quietly forward and at once received the mattero's full attention.

I had only ten paces to go when the bull turned and spotted me. He bellowed and then came at once in no uncertain manner. I sprinted forward, but as I did, Matador, alarmed by my sudden movement, began to move away. He was already at a canter when I drew along his off-side, with 1,400 pounds of fury pounding the earth only a yard or two behind. It was a left-handed mount, but fortunately one I had practised under less exacting circumstances. Grabbing a handful of mane in my right hand I vaulted and hit the saddle, hammering both spurs into the big bay's ribs.

Matador jumped as the sharp metal rowels stabbed him like an electric shock. Over my shoulder I saw a great white head with a horse's tail draped over one horn. I swung my whip in a wide arc and brought it down with all my strength on the pony's rump. The horns lunged again and mighty neck and shoulders followed through with a force that could drive the point like a spear through a horse's crotch.

But, they missed by a hair, and we were clear.

VII

SOME MECHANICAL PROBLEMS

When the round-ups were over I found myself faced with numerous requests to carry supplies to the outposts for the Indians who manned them. This was a job for the Land-Rover. We had two of these vehicles. One, Jimmy Brown kept jealously for his own use, but a second, which he called the 'Grey Ghost', stood in the workshop at the disposal of the cattle department. The 'Ghost' was the first vehicle ever to tread a tyre on Rupununi soil, and had been flown up piece by piece on small charter aircraft several years earlier. The remarkable thing about this was that Jimmy Brown, the man destined to reassemble it, had never put together anything in the mechanical line more complicated than a wooden-wheeled bullock wagon. At the time he must have been close to 50 years of age, yet suddenly a hitherto latent genius for the understanding of internal combustion engines came to light. Not only did he rebuild it, but also taught himself to drive, and within weeks had dual tracks blazed over most of the horse trails in the country. But the new form of transport was hardly without problems. Rocks drilled holes in the sump, and bearings lubricated with sand ground themselves into metal filings. Water at creek crossings got in the oil, the distributor, and on the spark plug points. Shock absorbers absorbed what they could for the first few miles and then collapsed exhausted. Springs were broken on almost every trip but fortunately the cracked were not discarded, they made excellent hinges for corral gates.

Jimmy Brown's amazing aptitude for putting mechanical equipment together was only surpassed by his incredible ability to smash it again shortly thereafter. Jimmy was a man of unbelievably strong constitution. Apart from a graveyard cough caused by the quantity of tobacco he smoked, which awoke the entire ranch every morning at 5 a.m., nothing ever went wrong with him. He could work like a mule on very little food and no doctor ever laid a finger on his pulse. Unfortunately he expected mechanical things to perform in the same way. Engines laboured on as best they could without oil, nuts and bolts gradually rattled themselves loose, and important parts eventually fell off. Every machine in Jimmy's charge seemed to defy the laws of engineering and continued to operate right up to the point of a major breakdown. Nothing, it appeared, ever had the nerve to quit with minor symptoms like dirty spark plugs, blown gaskets or disintegrated bearings. When an engine stopped it was a complete and total wreck; it had either thrown a connecting rod through the wall or pulverized the crankshaft.

Naught pleased Jimmy more than the last exhausted groan of a ruined motor. He would curse and swear, perhaps even kick it, but these were outward theatricals; within, Jimmy was as happy as a schoolboy with a new cricket bat. Now he could dismantle, examine, explore, radio Georgetown for new parts, belt bearings in and out of place with a 7-pound sledgehammer, skin his knuckles on obstinate bolts, then finally make them yield with a mighty heave on a pipe wrench. Above all, he would coat himself gloriously in oil from head to foot, and scatter nuts, bolts, gaskets, tools and related paraphernalia, the length and breadth of the workshop. Fifteen Indians would stand and watch in complete awe. These were his workmen, but they were rarely allowed to help. When Jimmy accepted the challenge of a broken-down engine he met it alone. Until he was finished his entire crew and whatever construction job they were working on would have to wait. Sometimes it took two days, sometimes two weeks, but one thing was certain; no matter how serious the injury James Brown would fix it, he was nothing short of a master.

On the other hand, I was the most inept mechanic that ever touched a wrench. The inner workings of four cylinder engines were to me a complete and insoluble mystery. It was not surprising that I found myself in difficulties almost every time I ventured out onto the road. Mr. Brown observed these breakdowns with a jaundiced eye and I gathered that he only enjoyed repairing vehicles that he himself had had the pleasure of smashing.

Only a short while before I had incurred his wrath as a result of an ill-fated expedition with a reluctant government agricultural officer. We were returning from the outpost at Raa Wau, some

40 miles west of Dadanawa on the Tacatu River. There had been an outbreak of foot and mouth disease across the frontier in Brazil, and I had driven the official out to our boundary so he could see for himself that no cattle in Guyana were infected. The 'Ghost' was puttering along at about 35 m.p.h. and we had just crossed the Sawara Wau river, which roughly marked the half-way point in the journey. It was not a very well-worn trail, one that Jimmy had only blazed a short while before. The grass was very long and obstacles in the road were difficult to see. Suddenly I saw what looked like a stump and swerved to avoid it. Had I continued on I would have found it to be only a harmless termite mound which would have disintegrated immediately. But I swerved, and discovered to our intense shock a very real and immovable stump concealed in the grass. The impact threw me against the steering wheel, but the official was catapulted forward and banged his head into the windshield. The Rupununi District was never considered a favourite area for a civil servant to be assigned to, and as we got out of the car to inspect the damage I overheard some scurrilous remarks about the people back in the Georgetown office.

The collision had been head on, shattering one of our headlights, which didn't work anyway, and putting a V-bend in our front bumper. I got back into the driver's seat, put the car in reverse and pulled away without any difficulty. "She's fine," I called to the official, who was dabbing a handkerchief against a minor cut on his forehead. "Let's go."

We drove on uneventfully for three or four miles until I began to detect an ominous bubbling sound above the rattle of the engine. Any extraneous noises that could be heard over the normal hammering of the 'Grey Ghost's' pistons against the cylinders was cause for immediate concern.

"Seems like we're boiling," the official observed.

"Probably nothing to worry about," I said reassuringly. "It's a damn hot day. There's water about a mile further on—we'll fill up the radiator there."

By the time we arrived at the spring the engine temperature gauge was on the verge of melting and a disturbing cloud of steam was forcing its way out of the bonnet.

We got out and lifted the hood. Both of us looked at the radiator cap. "Don't touch it for at least fifteen minutes," said the official.

"Oh, ridiculous," I countered, and produced a hammer and screwdriver from our bag of tools.

"Watch out now," I said, and placed the sharp point of the screwdriver against a notch on the rim of the radiator cap and lifted the hammer.

I've never seen the lid blow off an oil well, and only once or twice the cork go off a champagne bottle, but somewhere about half way between those is the top going off a red-hot radiator. The cap, I feel sure, went into orbit. It missed the officer's head by pure good fortune, and showered us both with steam as the entire remains of the radiator went into space like a geyser. Unfortunately, our seared eyebrows were not the full extent of the damage. Further inspection revealed a hopelessly bent fan and a broken drive belt, all sustained in our collision with the stump. It amounted to a major breakdown and we had no option but to walk the dozen or so miles left to Dadanawa.

When we arrived late that afternoon Jimmy greeted the news with characteristic crustiness. We enjoyed one of our longest conversations since my arrival at the ranch.

"Where the hell did you leave it?" he asked.

"About twelve miles up the trail," I replied.

"Son-of-a-bitch," he grunted. Then threw his tool box into the back of his Land-Rover and drove off alone towards it.

Mechanical breakdowns were not the only spectre to haunt the Rupununi driver. Bottomless quagmires crisscrossed the terrain, and shifting sands in the river beds moved under the wheels and settled the helpless machine on its belly. There it might lie for days, a useless chunk of metal, snickered at by passing Indians, stared at by cattle, shrieked at by angry birds. Finally the brawn of a dozen oxen would arrive and drag the monster to dry land. The twentieth century still could not live at Dadanawa without help from the past.

Assisted by four Indians I pushed the 'Grey Ghost' out of the workshop. We checked to see that the bare-essential items for survival were on board. A shovel, an axe, a machete, a wheel-locking nut spanner, and a set of other assorted wrenches, a screwdriver, hammer, two 4-foot lengths of inch board, and a jack—

the latter items indispensable when it came to lifting up a stuck vehicle in a muddy swamp.

Lovable as the old heap was, one of the irritating features of the 'Ghost' was its lack of a starting mechanism. It was also not the easiest of creatures to wind up with a crank, and had an appalling habit of backfiring on the third swing of the handle. As a result no one who had previously experienced a sprained or fractured wrist would volunteer for cranking duties. This left me with no alternative but to either crank it myself, or carry four boys on every trip so as to always have enough manpower available for a push start. Thus equipped we propelled the benevolent 'Ghost' to life, and loaded with supplies bounced and roared away down the sandy trail.

The 60-mile round trip had been strangely uneventful. All the merchandise was delivered and by mid-afternoon we were trundling home at a steady 30 miles per hour. About 5 miles from Dadanawa we took advantage of a descent down a steep hill to increase our speed. Suddenly there was a shout from one of the boys in the back and something appeared alongside of the vehicle. Before I had time to realize what it was the car took a terrific lurch to the side and slithered to a halt throwing one of the Indians from the back seat on top of me. We climbed out, and I was surprised to see that the object I had noticed was one of our wheels, still connected to part of the axle, lying several yards from the car in the side of the road. Fortunately though, despite the awesome appearance of things, this was by no means a hopeless breakdown, even for as inept a mechanic as myself. The cause of our misfortune is known technically as a slack wheel axle retainer. The 'Ghost' was one of an early line of Land-Rovers that were built with the rear axle in two halves, held together by the retainer. So after many miles, and much abuse, that part sometimes develops a weak point. Then, presto, whether your wheel-locking nuts are hammered on with a pipe wrench or not, when the retainer goes the wheel and half of the axle just fly off.

Luckily for me this was not the first time I had experienced this particular problem. Only a few months earlier, in the middle of the wet season, I had helped Jimmy ferry his Rover across the swollen Rupununi River on a raft. The raft was built out of 40-gallon sized petrol drums. Twenty-four were tied together with fencing wire and the top was floored in with hardwood boards. A moveable wooden ramp allowed the vehicle to be driven on board from the river bank. Then it was physically hauled to the other shore by about twenty men on the end of a heavy rope. It was really quite a simple procedure. In fact, the hardest part of all was invariably getting the rope itself across in the heavy current as this had to be dragged over in a dug-out canoe.

Well, when we got to the other bank, we ran out the loading ramp and Jimmy drove confidently forward, exuberantly puffing away at the inevitable cigarette. He was barely half way along it when his right rear wheel suddenly dropped off, complete with half of the axle, and the entire vehicle, including Jimmy, slid unceremoniously into the river. With all the manpower we had available the most serious damage turned out to be to Mr. Brown's cigarette. In no time at all we had lifted the machine back on to its three good legs, in what was fortunately shallow water, and the wheel and axle were slammed back into place. The interim repair, with cautious driving, was sufficient to get us back to base where a new retainer could be fitted.

So I was merely faced with an identical situation and all it required was to jack up the chassis. We pushed the axle back into its socket and once more were homeward bound. Or so we thought. We had scarcely travelled a hundred more yards before there was a horrendous tearing of metal, followed by a shriek from one of the back-seat riders. Simultaneously we were jarred to a sudden halt and once again toppled over on to our side. To my amazement I discovered that we had now lost our other rear wheel, which had not managed to spin itself clear. Instead, the end of the axle had been driven upwards through the floor of the vehicle, and been directly responsible for the cry of pain that emanated from the hapless Indian who had so unwisely chosen that spot to sit in. Despite our efforts we were quite unable to effect any kind of a field repair as this time things had been buckled and damaged. We had no alternative but to hike our way home, which for the injured man in our group, was a painful process.

When I got back to Dadanawa I found Jimmy had returned from his work on the corrals. He gave a grunt when I described my misfortunes with the 'Ghost', and announced that the next day we would drive in his Land-Rover to the frontier outpost of Lethem where we might find a spare part. Lethem was 62 miles north of the ranch. Its sole claim to existence was a police station, a district commissioner and two saloons. With the exception of the latter, the only other things there of interest were a couple of derelict vehicles from which certain parts, if one were lucky, could be cannibalized. It took us three hours to get there and another five searching through iron rubble to find a part that might do the job. Heartened by our good fortune we decided to pay a short visit to the saloon cum general store, before heading home.

The Lethem store was located on a bend in the trail that led up to the police station. A worse spot could never have been picked with even the most careful planning. It was downwind of the road and, as Lethem was situated in flat open savannah, blessed with a twenty-four-hour breeze of some 20 knots. Therefore, the passage of every vehicle, or even horse, kicked up a miniature sand storm which was then conveyed with unerring accuracy through the saloon front porch.

I followed Jimmy through a dilapidated door into the saloon. The light outside was so intense that the effect as we walked in was one of complete darkness. My feet stumbled over something soft, and seconds later as my eyes became adjusted, I saw it was a man sprawled on the floor.

"Drunken son of a bitch," Jimmy mumbled and stepped over him. There were three Indians leaning on the bar. They smiled with sloppy mouths and looked at us through dull red eyes. One of them poured three fingers width of rum into a glass and gulped it back. His head rocked forward with eyes screwed tightly shut, and muscles of cheek and neck tensed in pain as the liquid burned his throat. He stood there for a moment swaying, then reached towards us with the bottle in an inviting gesture. The manœuvre was too much for him. He fell like a tree, and bottle and glass shattered in his hands as they hit the concrete floor. Jimmy scarcely spared him a glance and I followed him through another doorway into a room that comprised the general store.

"I wanna buy a hat," said Jimmy, to the light-skinned shopkeeper who appeared to be a mixture of Amer-Indian and European blood.

The man said nothing but climbed a small ladder and removed four wide-brimmed bush hats from the shelf. He stepped down and handed one to Jimmy. Jimmy grasped it in two hands on either side of the brim and rammed it on his head. It stopped on the top rim of his spectacles and he wrenched it off as roughly as he'd put it on.

"Too damn big," he grumbled, and seized another. The next was too small and similarly discarded. The third apparently fit, or at least Jimmy reached that conclusion in a matter of seconds, and tossed a 5-dollar bill on the counter.

"Gimme the change in matches," he said.

The light-skinned man pushed a large pack of them towards him. There were twenty-four boxes in the pack, but as Jimmy's home made cigarettes had a habit of extinguishing themselves roughly once a minute, the supply would last little more than a week.

As we stepped back into the street Jimmy immediately removed his new hat and flung it as hard as he could to the ground. He then jumped on it with both feet and continued stomping until it was the colour of the road and the shape of a pancake. Finally, he picked it up and beat out the dust against his leg. Then, while holding it by the brim, he punched it with his other hand into some semblance of shape and stuck it on his head.

He looked at me for a moment and could not have helped but notice my look of stark amazement.

"The first thing yer do with a new hat is jump on it," he said, gruffly. "There ain't nothing more disgustin' than a new hat." He turned, walked over to the Land-Rover, got in and started the engine. I hurried after him, for I would not have been a bit surprised if he had driven away and left me there.

The road home was in very bad condition. Heavy rain from the previous wet season had left it full of holes and ditches. These become familiar after a while, but it takes several trips along a route to remember where they all are. Our visits to Lethem were infrequent, and therefore, not unexpectedly, the journey was a bumpy one, full of surprises.

The trail followed the foot of the Kanaku mountains for 20 miles and then, after rounding the westerly spur, cut across the flat infertile Macushi plain. We passed the dirty mud-walled hovels of the Indians who lend their name to this section of the savannah. The Macushis were the northern neighbours of the Wapishanas, but through ancient custom did their best to remain separate from each other. Finally, the last adobe homestead slipped by, and once again we were in Wapishana country. However, we would see none of these people until we crossed the Rupununi at Dadanawa. The hunting on the plains was poor and formed a natural no-mans-land some 20 miles wide between the two tribes.

There were 30 uneventful miles behind us and it was already late afternoon. The car sped down a steep hillside and the road took a slight curve into the rapidly setting sun. Jimmy squinted through his glasses as the red light and long shadows played tricks with the contours of the trail. He sat slumped in his seat, maintaining a steady forty, while I looked off to the side, away from the fierce sun, at a herd of grazing cattle. We came out of the curve almost at the foot of the hill, and there right across the trail was a jagged rut several inches deep. It was too late to brake or slam her into low. Jimmy had only a split-second, to either take the ditch head on or pull off the road. The latter was logical, after all, in most cases the only difference between the road and the surrounding countryside was a well-worn trail. When the trail got bumpy it was often smoother to leave it and drive along the edge.

If we hit the rut we would need a new set of springs, and in the momentary glance that Jimmy had of the situation it seemed that the gully stopped on the left-hand side of the road. Instinctively, he wrenched the wheel hard to the left, throwing the car in a wild two-wheel turn. Suddenly we were in long grass with trees dead ahead. Jimmy wrestled the wheel to the right pulling the car back towards the trail and the setting sun. A Curatella tree loomed out of the long shadows in our path. Jimmy swerved again. I barely glimpsed a sturdy pointed limb rushing with frightening speed towards the windshield. I flung myself sideways into the vacant middle seat under a shower of fragmented glass. There was a terrible jar and my back was flung against the metal dashboard and I grappled frantically with the gear handle for some kind of support. Metal tore like paper, and then, after three or four coughs and a final convulsive shudder, the motor died away and everything was strangely silent.

I rolled over on my back and looked up from the floor of the car. Leaves seemed to be growing everywhere in the cabin. I reached up and found myself pushing against a very solid piece of timber. The bough had come right through my half of the windshield, and was firmly burrowed into the backrest of the seat where I had been sitting. I wriggled out towards the driver's side, turning over onto my belly. The door was open and there was Jimmy, sitting in the grass rubbing his head. His spectacles were hanging on one ear. Carefully, he fitted them back into place and looked at me.

"Are you okay?" he enquired, which was not only something of a speech for a man so economical with words, but also the first time he had ever enquired about my health since I'd lived at the ranch.

"I'm fine, thanks, how are you?" I replied.

"Still in one piece," he said, and climbed to his feet.

I wormed my way out of the wreckage and joined Jimmy at the front of the vehicle to inspect the damage. We had hit virtually head on. Our left wing had been torn clear and stuck out to the side like the lame wing on a bird. Both headlights were shattered, and the bumper was firmly bent backwards against the left front tyre. I looked at the tree which had won such a convincing victory. It was barely 15 inches thick in the trunk, and decided that *Curatella americana*, hitherto considered as worthless timber, might yet have a future in the construction business.

We were 30 miles at least from Dadanawa. The nearest Indian help were the Macushis, a dozen miles or so back towards Lethem. If we couldn't fix it we had a long hike ahead of us. But Jimmy never abandoned his vehicle regardless of the damage. The first thing we had to do was get a lever and prise the bumper clear of the front wheel; then the two of us could push it back clear of the tree. I got our axe from the emergency tool kit and cut a pole from a tall slender tree that I eventually found a half mile from the scene of the accident. After a great deal of effort, we succeeded in bending the bumper approximately back to its original shape. The bough of the tree that had gone through the windshield was cut off at the base and dragged clear, together with our left wing and door. Jimmy picked up the crank handle and smashed away all the pieces of jagged glass left in the windshield until we were only left with the clean, empty frame.

"Should be nice and cool now!" said Jimmy; with a laugh. "Let's try the engine."

He kicked the starter, and to our immense surprise the engine started immediately. Unfortunately, it was accompanied almost at once by a strong smell of burning and blue smoke filtering out through the buckled hood. Jimmy switched her off, and I lifted the bonnet. Luckily it was only a minor problem. The impact had dislodged the air-filter cap which had fallen across the top of the battery terminals and was causing a short. We corrected it and the engine ran perfectly.

It was already dark, but we were soon on our way, with Jimmy at the controls, while I held a flashlight through the empty frame of the windshield on the road ahead. Behind us lay a pile of twisted metal, that would stand as a monument to our escapade for many years to come.

VIII

TRAIL DRIVE

Later that year Charlie unexpectedly announced his retirement. He was not in bad health, though he walked with a limp from numerous falls, and the index finger on his right hand was only partly usable. That finger was most important as it gave the final guiding touch to the lasso as it left the hand. Charlie's had been burnt to the bone when a steer jerked the rope unexpectedly through his grip, for rawhide cuts more quickly than a man's reflexes can tell him to release it.

His departure came as a great surprise to me for Charlie seemed proud of his stature as chief vaquero. But he had a wife and children, and, if the truth were known, he was probably tired of being away from them for long periods on the round-up. So he built himself a small adobe brick house, thatched with Mauritius palm, at the foot of the Kanaku Mountains, 20 miles from Dadanawa. He took up his bow and arrow, gun and fishing tackle, and overnight returned to the way of the Indian. The saddle and lariat that he handled so expertly were hung up forever.

I missed Charlie. I had learned most of what I knew by just watching him in action. Although we were not close friends, I felt that I had completely overcome the resentment that prevailed during my early months at the ranch. Since successfully breaking Cang, Charlie had always been ready with helpful advice. His knowledge of the 3,000 square miles over which the vast herd was scattered, was unequalled amongst the Wapishana tribe. He knew the name of every little creek and hill, and the quickest route to ride from anywhere to everywhere, day or night. There would no longer be the luxury of following his horse's tail through the pitch darkness along twisted wooded trails, knowing that his uncanny sense of direction would lead us flawlessly to our destination. Such experience was irreplaceable; his loss in terms of round-up efficiency would be felt for some time.

My respect for Charlie had brought about a secret ambition to become a round-up foreman myself. His resignation suddenly precipitated me into that position. I wasn't in any way officially promoted to the rank; there was no one there who was going to stand up and announce that Stan Brock had been elected foreman, or put a notice up on a board bearing an impressive seal. No single man was actually in charge of Dadanawa. The chairman of the board had never even heard of Charlie, much less that he had retired. So all at once there was a vacancy, and, as there didn't seem to be any opposition, I just took up the reins.

Of course, there were many Indians who could have filled the post more adequately than I, but probably few would have been interested to do so. Charlie was unique, at that time being the only Wapishana who held a steady job. Work for wages was not the Indian way of life, and few would stay longer than a single round-up season. In fact, probably only 2 or 3 per cent of the tribe hired out at all. Some bled balata, a form of rubber latex found in the surrounding forest, but most were content with village life. Every year each family cut and burned an acre or so of virgin forest, and planted manioc, maize, bananas and watermelon in the sandy soil. When the crops were ready, they lived on the manioc and fruit, and sold the maize at Dadanawa to be used as horse feed. Its worth was bartered for cloth, ammunition, cooking pots, axes, knives and fish hooks. Armed for another year the Indian spent it in leisurely pursuit of fish and game. What did he care about the prestigious position of 'capitaso' that Charlie held at the white man's ranch.

My first major assignment came in September shortly after the rains. A herd of 1,000 head was required in Georgetown. This was several weeks earlier than usual, for the trail in the forest was still wet and even light rainfall could suddenly flood the many streams and rivers that had to be crossed.

The branding season was not yet underway, but fortunately we were holding 1,500 Longhorn steers in a valley between the Rupununi River and the Kanaku Mountains. They had been put there when the river was dry in April and held by horsemen until it flooded in May. When the water rose they were imprisoned by natural boundaries, and by the time it fell again in August, were already settled on the new range and fattening. Rather than call a major round-up I recruited twenty men, cut 1,000 of the best steers from the herd and left on the 300-mile journey to the coast.

Any romantic conceptions that one might have about trail drives disappear by mid afternoon of the first day. The cattle though wild and ready to explode in stampede at the slightest provocation, have simmered down to a reluctant walk. The sun is settling down on the Brazilian side of the sky, beating fiercely in under the brims of the vaqueros' sombreros. The air is still, and a giant column of brown dust envelopes the riders and their herd. Ponies plod unwillingly in the Longhorn's wake, nostrils clogged with sand and heads sagging to within inches of the ground. Here and there they pause to snatch at a tuft of wiry range grass, and the ill-tempered cowboy jerks angrily at the bit and hacks him with a spur for breaking the tempo of his stride.

Water holes across the plains were few and far between. They were puddled by the hooves of visiting cattle, and littered with their dung and salted with their urine. On a good day a cowboy could expect three opportunities to slake his thirst, at sunup before breaking camp, at the midday halt, and later at sundown. On a bad day there would be no water between dawn and dusk, and that thick and yellow with the consistency of salad dressing.

The herd was drifted along at about 3 miles per hour at a rate of 10 to 15 miles a day. At noon they were rested for two hours, and arrived at the evening camp site an hour or two before sundown. Prior to dark they were allowed to spread out and graze over a half square mile, while vaqueros patrolled the perimeter on horseback, turning back those inclined to stray. The Longhorn has a homing instinct surpassed only by the carrier pigeon. Steers have been known to escape 100 miles from their home range and return to it within two days.

As dusk approaches the herd is gathered together and held by the riders in a tight circle. At the same time, numerous fires are lit around them, a large supply of wood having been gathered by a section of the men during the afternoon. At first the cattle mill uncomfortably but gradually settle, partly soothed by the constant calling of the vaqueros. Within a few hours most of the animals bed down, and the men share watches to guard them. The slightest unusual sound, or smell of a passing puma or jaguar, could cause an instant stampede. Then every animal would be on its feet in a flash, racing berserk over fire and sleeping vaqueros, to be lost in the darkness in seconds.

Ahead of us lay nine days on horseback across the plains to the edge of the forest. There we would leave our ponies and drive the herd on foot down a narrow trail cut through the virgin forest. This phase of the journey would take sixteen days. It was the hardest and most dangerous, crossing mountainous terrain and many rivers. Losses from the herd could be high. A successful drive to the coast was measured by the extent of them.

I took two horses, Butterball and Cang. The Indians had only one apiece, but they could take things easy on the return trip. Branding season was due to begin in six weeks, and I would have to ride hard on the way back in order to be in time for it.

The first three days passed uneventfully. The steers were strong and it was all we could do to hold them down to a fast walk. At this stage they had little interest in grazing and at the slightest excuse would break into a run. Their exuberance made driving easy and most of us were able to ride either in front of the herd or along the flanks, relatively free from dust.

At night I took the first watch between sundown and twelve o'clock. The cattle were most restless during that period, but being constantly occupied in guarding them helped the time to pass more quickly. At midnight the relief took over and I was able to fall into my hammock slung between nearby trees and sleep with few interruptions until 4.30 a.m. To prevent our horses from straying, we hobbled three of their legs together. A small plaited-leather hobble linked the two front hooves 6 inches apart, while one some 30 inches in length tied the near forefoot to the near hind. Even so, despite this restriction, many of the horses became so adept at moving that they would travel a mile or more during the night. This meant that the owner had to wake up at least once every hour to find his horse and re-hobble it nearby. A pony that was a bad strayer was often linked with the longer hobble diagonally across from the near fore to the off hind, and this method greatly decreased its ability to move. It was important that our horses remain no further than 20 yards from our hammocks so they could be mounted swiftly in the event of a stampede.

On the morning of the fourth day, we rounded the westerly spur of the Kanakus and headed due north across Macushi Indian country. That evening we would corral the herd at Manare, a small horse ranch which belonged to Dadanawa. It would be a welcome change from riding herd at night.

The distance to Manare was not great and by mid-afternoon we were within 3 miles of the house. Leaving the herd to graze amongst the tall grass of a Mauritius palm swamp, I rode ahead with one of the Indians to inspect the corral.

Shortly after departing from the herd we found it necessary to ride through a patch of tall timber. We had only just entered when all at once Cang gave a gigantic spring to the left. He didn't turn his body at all but just leapt sideways, almost leaving me behind. Caught off-balance, I was forced to grip him tightly with both legs, and as a result stuck him hard with one of my spurs. The answer was an indignant snort and Cang plunged ahead in a series of jarring crow hops. By this time I was already in trouble and in order to stay on board had my calves locked tightly into his ribs, the spurs tickling him unavoidably at every jump. He swerved to the right, and then to the left, squealing madly, with his head lost between his front hooves. In a moment we were completely out of control and heading for the stout overhanging limb of a Curatella tree. Cang couldn't see it, and I knew it was going to wipe me off his back about belt high.

There was only one thing to do. A desperate manœuvre that would pull him in a ninety-degree turn to the right and miss the branch. I lifted my left foot and rammed him hard with the spur in the point of the shoulder, and simultaneously dragged with all my force on his mouth with the right rein. He spun around in his own length, thrashed wildly with his right front foot while his head was down, and caught it in the reins. Cang tripped, hesitated for a moment balanced on his nose, then crashed forward, rolling on his side with me still in the saddle. Luckily I was unhurt, and wriggled quickly out from under him, grabbing my halter rope in the process lest he should jump up and escape. After staggering to my feet, I found my Indian companion highly amused with the entire proceedings. He laughed and pointed up into one of the nearby trees. About 20 feet above the ground, carefully positioned in a fork, was a very large giant ant bear!

It had been the sudden sight or smell of the animal that had frightened the unpredictable Cang; but the strange thing was it had no business being up in a tree in the first place. Unlike their small, prehensile-tailed cousins, the tamanduas, the climbing capabilities of the giant variety are extremely mediocre. Furthermore, there were no termite nests growing on the tree to have warranted the phenomena anyway. It was a full-grown specimen, weighing about 120 pounds, and measuring some 7 feet from the tip of its tube-like nose to the end of its bushy tail. The ant bear is perhaps the most bizarre looking of all the creatures on the South American continent. It is covered in long wiry hair, which on seven-eighths of the body is jet black. Only the forearms are white, running back in a diagonal line to the withers. With its shaggy tail waving in the air like a witch's broom, small wonder it alarmed the wily old Cang.

Manare was comprised of one mud-walled house, a corral and a very attractive clear-water stream meandering quietly by it. As our horses approached, we were rushed by several mangy and ferocious-looking dogs who barked belligerently at our heels. Their noise soon brought the foreman, an elderly Brazilian, storming from a doorway carrying a stock whip. He recognized us, and immediately began to curse and lash at the dogs unmercifully. He caught one cross the back and it ran howling off towards the river bank followed swiftly by the others.

"*Bon dia,*" he said.

"*Bon dia,*" I replied, then dismounted and shook his hand.

"*Tros munto boi* [have you brought a big herd]?" he enquired.

"*Um bucardo* [quite a few]," I answered, "*Mas o menso mille reis* [a thousand, give or take a few]."

"*Entra,*" the Brazilian said amiably, waving towards the doorway. "*Vamas toma un cafe sino* [let's have a little coffee]."

After enjoying two thimble-sized cups of black Brazilian coffee, scented with aniseed, we inspected the corral. It was of typical sturdy construction and the gates were in good repair. Satisfied, and with dusk approaching, I remounted Cang and rode back towards the herd.

A low cloud of dust hung over the rolling terrain to the south, and smoke from the chain of trail fires rose in great pillars beyond it. Suddenly, a black line appeared on the crest of a hill, and an instant later turned into a thronging sea of hustling Longhorns as they poured over onto the open plain.

Two outriders sped down on each side of the surging mob and turned the leading steers who were strung out several hundred yards ahead of the main herd. They were close now, and I could hear the shouts of the vaqueros and the hard, sharp crack of the stock whips.

A fire sprang to life in the grass alongside the cattle, and flames licked hungrily at the sun-scorched ground. Soon its smoke would tower skywards and more blackened earth would follow in the vaqueros stride. The habit of the South American plains Indian to indiscriminately burn grass wherever he goes is a phenomena that defies both explanation and prevention.

"*Wat-ena wun* [it's going to rain]," said my companion in Wapishana.

I followed his gaze to the eastern horizon. A jet-black belt of cloud was creeping slowly up over the Kanaku Mountains. It was split intermittently by cruel jagged forks of lightning, followed minutes later by the distant rumbling roll of thunder.

"*Moonaperi* [it's a long way off]," I mused. "*Peu-nadon y duptan* [maybe it will pass by]."

"*Peu-nadon,*" the Indian agreed tentatively, but I could tell by his tone that he did not think as I did that the storm would miss us.

However, it was certainly an hour or two distant. Plenty of time to corral the herd at sundown. Thankfully tonight we would have a roof over our heads.

By dark the storm seemed little nearer. The herd was safely in the corral, and large fires had been built in front of the entrance. The gates were made of heavy poles which slipped horizontally through drilled holes in upright timbers. Each pole weighed about 100 pounds, and there were six, one above the other about a foot apart. As an added precaution, one pole was placed diagonally across the six, and then all were lashed together with lariats. Wise old steers had been known to work their horns against gate rails during the night and open them, allowing the entire herd to escape. Even this measure was not considered adequate on a trail drive, and the vaqueros were required to sling their hammocks around

(*above*) A giant armadillo drags the author

(*below*) A tapir caught in a heavy game net

(*left*) Tiny, the tapir. (*right*) Big Jacko and Tiny were the best of friends. (*below*) Irene with Tiny and the two young tapirs

A prehensile-tailed porcupine

the perimeter of the corral, ready to thwart a possible stampede.

I had my own hammock tied to the left of the gate and Butterball was hobbled only a few yards away. It was 9 o'clock and I had just dozed off to sleep when the Brazilian foreman shook my shoulder and invited me to the house for dinner. Actually I had already roasted some dry beef with the Indians, but not wanting to appear inhospitable, thanked him and accepted the invitation.

The interior of the house was darker than the outside. There was no lamp and only a small wick soaked in paraffin flickered with one candle power on the wooden table. I sat down on a bench and a few moments later a silent figure slid dimly into the room and set two bowls on the table, a tin plate, and a spoon. The figure who I took to be the wife, departed, and the Brazilian informed me that dinner was ready. Apparently he was not going to join me himself, for he sat down on a stool in the corner and without further word engrossed himself in a cigarette, pausing only occasionally to spit on the floor.

I peered into the two bowls and ascertained eventually by feel with the spoon that one contained ground roast beef. The second was filled with farine, the grit-like staple made from cassava that had become a daily dish since I first arrived at Dadanawa.

After eating four or five spoonfuls of beef I felt an unusual burning sensation about the fingers of my right hand, lips and cheeks. I continued eating, but the burning became steadily worse and finally got to the inside of my mouth as well. I thought at first that the food contained a large amount of red pepper, but then quickly discounted this theory as the burning did not affect my throat and stomach as hot pepper usually did. By the time another dozen spoonfuls had elapsed, the irritation had extended to my chin and ears. Unable to stand it any longer, I reached for the candle and held it close to my plate. The beef was a seething mass of red ants, which I suspected must have found their way to the bowl via the legs of the kitchen table.

I got up, thanked my host for the meal, and groped my way towards the door.

The storm must have crept up like a jaguar stalking a deer. As I stepped outside, there was an ear-splitting crash of thunder, followed by a sheet of lightning that went to ground only 100 yards from the corral. I was so startled I fell back against the mud wall and had it not been there, probably would have fallen.

There was a frantic shouting of men, "Whoah—whoah—whoah," and then their voices were drowned in another thunderous roll. But this did not come from the heavens. A chain reaction of fear flashed through the dormant herd, and all, as one, surged forward in a blind, panic-stricken stampede. A thousand Longhorns were on their feet and running as if on the open range, forgetting the timber walls that stood in their path.

A moment later came the deadly explosion of splitting wood and crushing bone as a solid wall of steers hit the side of the corral with one berserk united momentum.

The uprights, 10 feet high, with roots planted 4 feet into the solid sun-baked earth, and made of wood so hard and heavy that the surface can turn the blade of an axe, were snapped off like matchsticks. Forty steers in the lead at the time of impact were crushed to death by the sheer weight of their numbers following behind. The horrible bellowing of the dying and maimed was drowned under the pounding of hooves as the survivors poured through the gap.

I started to run towards the corral with the first sound of the stampede. Then, as they smashed free, the two fires in front of the main gate were kicked asunder. I knew at once the herd was coming straight for me.

Turning, I sprinted back to the house. As I made the corner of the building a sheet of lightning lit up the oncoming mob. They were only yards away, a wild-eyed uncontrollable wall of death. I flattened myself against the mud bricks and they roared by. Seconds later the rain came.

"*Kademen wom kowarrun* [quickly, the horses]," I yelled, repeating the order for horses as I frantically sought a bridle in the darkness and raced for Butterball.

The faithful old buckskin was with the first bunch of ponies I found. He didn't snort, balk, or even flinch as I threw the reins over his head and rammed the bit between his teeth. I pulled off the hobbles, vaulted on his back and he was away.

The herd had headed north and would probably split up after a few miles. I lay low over Butterball's neck to avoid the trees, and gave him his head.

Frequent sheets of lightning helped me to keep my bearings

and look out for cattle ahead. Then all at once, out of the driving rain came another vaquero. I shouted across to him and we both pulled up.

Ny-una d'yen [where are they]?" I asked.

"*Um tika pudumut* [I saw them]," he replied excitedly. "*Ditty y makun* [They went that way]."

He waved his hand in the direction he'd seen the herd running a short while before and we galloped after them.

Butterball was running powerfully under me. The sweat from his broad back had soaked through my trousers and was beginning to burn, but it was a relief to know that I was well mounted.

A mile further on we passed two straggling steers, the main herd could not be far ahead. Sure enough, the next flash of lightning revealed them. They were no longer running in a tight bunch, but strung out one behind the other in what seemed a never ending line. I urged the buckskin on, and gradually we began to overtake them.

It soon became obvious that we were chasing about 300 head. The herd must have split up a mile or two out of the corral. Our only course was to stick with these and hope we could hold them. There were eighteen other men scattered over the countryside and, with luck, they might pick up the rest.

Finally we drew alongside the leaders and tried to turn them. Riding only a few feet in front of their horns we screamed in their faces until we were hoarse, but with little effect.

Mile after mile Butterball and the Indian's pony ran doggedly on. At last we wore the Longhorns down to a slow but determined trot. Luckily the rain began to ease off and the heavy bank of cumulo-nimbus drifted away to the west. A last quarter of moon peeped above the distant horizon and our visibility began to improve. Suddenly a rocky hill loomed into view 100 yards ahead. We steered the herd straight into the side of it and forced them to a halt.

It must have been about one o'clock, judging by the height of the moon. I was glad when the rain stopped, but now a bitter cold wind had struck up. I stood on the lee side of Butterball and pressed my chilled body against his warm skin. There was nothing we could do but stand by the cattle until daylight.

At 5.30 it was light enough to see. To our surprise we found another group of steers, numbering some 200, held by four other boys not half a mile from our position. We joined the herds together and drove them back towards the corral, collecting thirty-five more on the way.

An incredible scene of devastation faced us when we arrived. A strip of corral fifty yards long had been smashed to smithereens. Dead steers were piled up three and four high amongst the debris, while others lay moaning or stood with fractured dangling limbs.

The full extent of the horror only revealed itself when we began to clear the broken carcasses away. Underneath we found my hammock and a few feet from it, the bloodied white cotton of another. The owner had slung it between the uprights of the main gate. He was still in it, and perhaps had never seen the stampede, or even heard it, as the murderous tide of 4,000 hooves swept over him.

It took us three days to track and round up a total of 915 of the original herd. Fortunately, except for those that died, the remaining strays would quickly find their way back to their home range, for we were still little more than 60 miles from Dadanawa.

The following day, nineteen weary vaqueros headed the Longhorns north, once more towards the coast.

IX

THROUGH THE FOREST

It was fortunate that we had no more trouble between Manare and the edge of the forest, for by the time we arrived there five days later our ponies were spent. Four of the men abandoned theirs on the third afternoon and completed the last 20 miles on foot carrying saddles and hammocks on their backs. I rode Cang exclusively during this section of the journey as Butterball was beginning to look tired. This I put down to the hard riding the night of the stampede and, of course, his age which was at least five years more than the cantankerous grey.

The trail entered the forest at the eastern shoulder of the Pakaraima Mountains, at a place called Surama. The Pakaraimas at this point were similar in height and appearance to the Kanakus, but destined to spread westward into Brazil and Venezuela, eventually to lead to the towering Andes.

Surama consisted of a corral and roughly constructed shelter of palm leaves. Here we would leave the horses with two men and penetrate the jungle on foot. The trail averaged about 20 feet in width and, being cut through primary forest, was lined on both sides with tall trees and dense vegetation with little or no grass. If the herd stampeded, any mounted pursuit would be impossible. Therefore there was no value whatsoever in taking our horses any further.

Due to the narrowness of the trail we divided the herd into two parts. Thus, myself and seven men, our supplies on our backs, and 458 steers, entered the forest. The remainder of the Indians and animals were to follow two days later. Nearly 200 miles ahead lay the Berbice River, and a steamer that would ferry the herd from there to the coast. If everything went according to plan it would take us sixteen days to get there.

It must have been a strange experience for a herd of Longhorns, born and trekked over open plains, to be suddenly funnelled into an evergreen tunnel. Their reaction was one of obvious fear. Foot-weary as they were they broke into a run, hoping no doubt that the trail would end and once again they would break out on to familiar grassland. Four of us ran ahead of them carrying branches in our hands which we waved menacingly in the leaders faces whenever they threatened to overtake us. Two other men were stationed in the middle of the column, in a further effort to slow them down. The remaining two followed at the rear.

For five miles we moved at a steady jog. Steep and rocky terrain slowed the herd down on the uphill stretches but made them exceedingly difficult to control as they plunged over the ridges and into the valleys. We were each carrying about 60 pounds on our backs and whenever we could belabour the heads of the lead steers and persuade them to stop, brutal as it may sound we did so. Finally they submitted, and settled down to the long walk.

After a dozen miles we came to a clearing in the forest. It measured about 80 yards long by 40 wide and was roughly fenced with poles. Lush green grass stood knee high and the steers dragged it out in ravenous mouthfuls. A small creek ran across one corner. Similar clearings to this stood at strategic points throughout the trail. Some would have grass, others would not. Whatever the case the herd would spend each night in one of them. To prevent animals straying either forward or backward along the trail four of us slung our hammocks at each end of the clearing. It was a familiar occurrence to be awakened in the middle of the night with the ring-eyed face of a Longhorn peering into one's own. Tired as they were they never gave up their efforts to escape.

By the third day the herd marched along like an orderly column of troops, five and six abreast. That morning I had awoken with an uncomfortable swelling beneath my left eye, which had mysteriously developed during the night. To take my mind off it, and also because we were running low on fresh meat, I strode on ahead in search of game.

The spoor of red and grey brocket deer were common. Both animals were exceedingly timid and never presented more than a fleeting target as they darted across the trail or melted cautiously away into the shadowy undergrowth. In every case I held my fire. I had no desire to attempt anything but a certain shot, as there was no time to spend in tracking a wounded animal deep into the forest.

Overhead the monkeys were legion; slow moving, grunting, belching, bearded red howlers; and the incredibly agile black spider monkeys, with their long prehensile tails and fantastically co-ordinated limbs that made them look like giant arboreal tarantulas. Little yellow sakiwinkis chattered and bounced from tree to tree in the hundreds, and here and there a tiny pair of marmosets flitted across enormous gaps as if their feet were fitted with coiled springs. There were white-faced sakis, and bearded sakis, with faces like great Mongolian warriors and woolly, poker-straight tails carried erect like proud battle flags. Capuchins ran across the trail by the dozen and were one of the few primates that dared to descend to the forest floor. They barked like dogs, and the Wapishana word for the species is very appropriately 'Wow'.

Occasionally I crossed the tracks of tapir, the largest animal to be found in the South American jungle. They left a three-toed, maple-leaf shaped print, some 6 inches in length and 5 inches broad. Because of their weight, which in a large specimen could be 500 pounds, the spoor was relatively simple to follow. However, there was more meat than we needed on these inoffensive relatives of the rhinoceros, and I left them be.

After some three hours I was still unsuccessful in my quest for a deer and decided to settle for an aguti. The aguti is a large rodent weighing about 15 pounds. Its body is covered in long hair which is mottled greenish-brown over the shoulders, and gradually turns to vivid orange at the rump. The ears are short and square and the tail is a bare stump little more than an inch in length.

Unfortunately for agutis they were born with two drawbacks. One they are excellent eating, and two they are endowed with an inquisitive nature. They are a favourite prey of all the cat family and when attacked, if they have any breath left in their lungs before they die, emit a high-pitched repeated whistle. This cry of alarm for some extraordinary reason not only attracts competitive predators but also other curious agutis. Hence more than one has unwittingly launched himself into the very jaws of destruction; and more than one feline has had to fight off another in defence of its prey.

I had learned from the Indians that a green leaf rolled into a cone and sucked through with tightly pursed lips so closely simulated this distress call that any nearby aguti found it almost impossible to resist. Thus armed, I positioned myself behind a large tree, and, after five minutes of calling was once again on my way, carrying a plump aguti strung over my shoulder with a piece of vine.

For eight hungry men one 15-pound rodent, even eaten complete with skin, is a meagre meal. Therefore, I decided to bag one more before rejoining the herd.

I selected a spot on top of a ridge where there was ample cover behind large boulders and the sound of my calling would drift effectively into the valley below. I sat down with a rock as a backrest, rolled a crisp new leaf, and, with rifle cocked across my knees, began to whistle.

After several minutes I heard the unmistakable hop-skippedy foot patter of an aguti approaching through the fallen leaves. It stopped, and I could visualize him with head cocked intently, the two chisel-like front teeth bared and one foot poised undecidedly about his next step. I whistled once more in a five-second burst. Almost immediately the footsteps began again. Twenty yards away they seemed. Soon he would appear behind one of the boulders. I raised the rifle and glanced with both eyes open along the barrel.

Suddenly the delicate pattering stopped. A moment later there was a snort of alarm, followed by a hard thumping of hind feet and breaking undergrowth. The aguti was gone.

I lowered the rifle and looked down for the leaf which had dropped from my lips. The quarry had undoubtedly caught my wind, but perhaps there was a second animal close by. It was worth a second try. I raised the leaf and looked up.

Ten yards away at the base of one of the boulders, staring me right in the eye, was a jaguar. All at once my mouth was as dry as a stick, and the back of my neck tingled apprehensively. Though no larger nor more deadly than other jaguars that I had seen, there was something terrifyingly sinister and ominous about this creature. There was little light filtering through the tall trees and the dim outline of his body blended into the dark background. He was a great sinewy chunk of muscle with green hypnotic eyes which penetrated, chilled, frightened and overpowered. He was crouched forward so that his chin was only a few inches above the rocky ground. The weight of his body was poised over bent bulging forearms, and hindquarters were tucked under his belly ready to impart their full power into an electrifying spring.

His arrival did credit indeed to my imitation of an aimal in distress. But at that point the last thing I wanted to be mistaken for was an injured aguti. With the slightest movement of my hand I checked that the safety catch on the rifle was off, and curled my finger reassuringly around the trigger. I had no desire to shoot the animal, and anyway a dead jaguar in the middle of the trail was more than likely to stampede the herd, which could not be more than half an hour behind me. The cat made a simple shot where he stood, but he was less than a jump and a half away. If he decided to attack he would be on top of me before I had time to raise the gun. The question was whether I should remain perfectly still and hope that he would decide I was a non-edible species; or try and kill him before he decided otherwise.

There is a limit to how long two animals can stare at each other without either flinching or blinking, and the human species, if I am an average example, is probably the least adept at this type of control. The sweat trickled from my forehead into my eyes and they burned. Then, to my horror, they blurred. The image before me suddenly began to dissolve. I had to do something. I had to shoot.

In one desperate movement I hit my right eye with the back of my hand and threw up the rifle. My finger came back on the trigger and the 30.06 shattered the jungle stillness. A hundred and eighty grains of copper and lead whined off a boulder directly behind where the big cat had stood and spent the rest of its energy harmlessly amongst the trees. I barely saw the target as it leapt sideways into the undergrowth. It had been moving even before the rifle got to my shoulder. Had he chosen to fight instead of escape I would never have stopped him.

When I awoke next morning my left eye was almost completely closed with the swelling underneath extremely painful and the size of a small egg. I concluded that the problem was an abscess and the sooner lanced the better.

Minor surgery on readily accessible parts of my own body I normally performed myself. However, on one occasion, months before, while sitting down on a log under a heavy load, I managed to deeply implant a thorn in that area where one has great difficulty in seeing without a complicated system of mirrors. So the operation was ably carried out by one of the Indians with the sharp but rusty blade of a sheath knife. As my own back was the sole vehicle for carrying supplies, luxuries such as a mirror had been left behind on this trip. I had long since learned to shave by feel with an open razor that I stropped across the thigh of my denim trousers. Therefore, without a mirror I was reluctant to make any exploratory incisions so close to my eye and, albeit hesitantly, summoned my second in command to do the job.

José was a Wapishana with short stubby fingers like gnarled caimbé bark which he used on my swollen cheek like a pair of pliers. I sat on a tree stump while he pushed, prodded and twisted the lump in all directions, peering at it with deep-brown eyes only a few inches from mine. Suddenly he jerked upright with his mouth open in a wide smile comprised of only three nicotine-coloured teeth.

"Boss," he exploded. "*Auna toomo oori, cooch aaga* [that's not an abcess, it's a cooch]!"

I looked at him in amazement. What I thought was an abcess, was in fact a cooch, the Indian name for the larvae of a tropical species of bot-fly. When they mature, these maggots can be an inch and a half long and the thickness of a man's little finger. They are very common among all the mammals of the forest, including the jaguar, and manifest themselves in large swellings over the animal's hide. They live just beneath the skin and have a tiny orifice through which they breathe. The pointed nose of the larvae reveals itself periodically at the orifice opening to gulp air. José's eyes had spotted it, and the reason for such surprise at the discovery was that it is extremely rare for the cooch to select man as its host.

"*Napen wom sootan* [how will we take it out]?" I asked, in a voice that I'm sure conveyed some nervousness.

"*Soom ed,*" he replied firmly.

I looked at José for a long moment, wondering just how he proposed to extract a wormlike creature from my face with a cigarette. Unperturbed, he took a plug of black tobacco from his back pocket and proceeded to scrape off a portion with his knife. A few minutes later he had rolled it between a paper and lit it up with a brand from the fire.

By this time the other Indians had gathered around showing great interest in my predicament and each insisting on giving my swelling a squeeze. As the larvae's head appeared they chattered excitedly poking it with their fingers while nodding approval of José's accurate diagnosis. The herd, already half an hour late for the start of the day's trek, was temporarily forgotten.

José took a long deep draw at the cigarette, filling mouth and lungs with the strong smoke. Then he opened the palm of his hand, placed it barely an inch from his lips, and blew the smoke into it as hard as he could. The ritual was repeated perhaps twenty times, gaining rapidly in tempo towards the end until José was drawing a puff from the cigarette and blowing the smoke in his hand with the pace of an accelerating locomotive.

Finally the cigarette was finished and the butt thrown aside. Then I saw in the Indian's palm a black pool of tar and nicotine, barely enough to cover a match head. José drew his knife and scraped the sticky substance up onto the point and moved towards me. I stood there apprehensively wondering what would happen next as he deftly smeared the tar off the knife point onto the swelling covering the orifice.

"Plaster," said José in curt English.

I produced a roll of adhesive tape, which formed exactly half of our medical kit, and handed it to him. José cut a 2-inch length and stuck it over the swelling, pushing it down firmly with his rough fingers. He stood back and looked at me with great satisfaction, and motioned with his arm in the direction that the sun would set.

"*Wo-cheepenin wom sootan* [this evening we'll take it out]," he said.

That evening after the day's drive, and with the cattle safely bedded down in the next clearing, José announced that it was time to remove the cooch.

I was instructed to lay down on my back and he knelt astride my chest with his weight on it. Another Indian positioned himself alongside my head. José partially clenched his fists and placed a row of knuckles facing each other on either side of the swelling. Then he instructed the second Indian to carefully remove the piece of plaster.

I felt the plaster begin to pull at the tender swelling as it was slowly peeled back. Then there was a pause.

"*Um ticup irroowai* [I can see the head]," exclaimed the second Indian excitedly.

José calmly ordered his assistant to take hold of the now exposed larvae's head, which had become stuck to the plaster during its stifled efforts to breathe through the morass of nicotine. The boy caught it between his finger nails and announced he was ready.

Without further ado José drove his two fists together and flattened the swelling as if it were a piece of dough. An excruciating pain shot through my face and it was all over. I opened my eyes and the two Indians were sitting back on their heels roaring with laughter. Still attached to the fingernails of one was a white curly cigar-shaped maggot, perfectly intact. It was a little over an inch in length and rather thicker than a cigarette.

So efficient was that jungle surgery that by the next morning the only marks on my face were a row of bruises from an over zealous, but effective pair of Indian fists.

Early next morning we arrived at the Essequibo. The Essequibo is the largest river in Guyana and bisects the country north and south for some 620 miles. At our crossing point it was about three-quarters of a mile wide and made up of a number of turbulent rapids rushing between several densely forested islands.

There was a small Indian settlement nearby, and a large pontoon powered by two outboard motors was maintained by the villagers for the purpose of ferrying cattle across.

The herd was forced into a corral on the river bank and then driven through a narrow chute onto the raft. It took about forty animals at a time and there was a raised barrier of poles surrounding the raft to prevent cattle from jumping out.

We got the herd across in eleven loads; an operation which took more than half the day. Everything went well until the last load, when about half way over the pontoon struck a submerged rock. This was neither serious nor uncommon, but the jar jerked the vessel to a sudden halt. The steers became alarmed as two men ran along the side of the raft to pole the bow clear. One of them panicked and leapt at the barrier. The heavy animal struck the top rail, smashed it to pieces and toppled overboard into the current. There was no time to get a lasso ready before the rapids had swept it far out of reach. We watched in silence as the steer waded ashore onto one of the nearby islands. It turned, looked at us, and tossed its massive horns defiantly, before trotting off into the thick undergrowth.

This was the first animal we had lost, and there was not the remotest chance of recovering it from the island. Yet to leave it there would mean for the animal either gradual starvation, or a swim to the mainland through murderous rapids. Even if it survived the latter it would fall prey to the jaguars during the first night of freedom.

Fortunately for us there was a solution. Rations were low and one of the weaker steers was normally slaughtered at this stage of the journey for meat. After seeing the remainder of the load safely ferried across three of us paddled back to the island in a dugout, to do what had to be done.

As we stepped ashore I fed a 220 grain solid into the breech of my 30.06 and set the safety. The level of the island was barely above high-water mark and the footing was soggy. To further complicate matters the thick vegetation was tangled together with a particularly tough-fibred slender-leafed plant; aptly called, because of its knife-like edge, razor grass. The efforts of three men trying to penetrate this thicket sounded rather like a work party of elephants dragging timber through an Indian rain forest. It was foolish to suppose that we could sneak up on a renegade steer under those conditions, and at best our visibility was only a dozen feet. Realizing that our efforts would be both futile and dangerous I signalled my companions to return to the shoreline. After a brief discussion I sent them to the upwind end of the island, with instructions to make as much commotion as possible, by shouting and throwing rocks into the undergrowth. As the island was barely 40 yards by 20 in area I felt certain that the animal would soon present itself for a shot.

The opportunity came quicker than I expected. I was moving along the shoreline and the men had not yet begun to shout when suddenly the animal broke cover but a few yards from me in a determined charge. Fortunately my thumb was on the safety and at the first hint I pushed it forward. There was no time to aim and I flung myself backwards into the water, firing into the brown of him as I fell. The bullet hit and turned him, for as he skidded sideways one of his front hooves slipped on the wet ground and struck me in the ankle as I jumped away. When my head broke the surface he had already disappeared back into the jungle.

I called to the boys and told them that the animal was hit and to refrain from shouting; hoping that a moan or stumble might give away its position. After fifteen minutes we had heard nothing, so I told them to make a commotion and throw more stones. Again there was no response from the old Longhorn.

After waiting impatiently for several more minutes I decided to follow the spoor into the thicket. I did this with the greatest caution moving only a few feet at a time. Shortly I encountered a heavy blood trail containing much froth. The sign denoted a solid hit in the lungs and I felt sure the animal was already dead. This opinion had scarcely time to form before there was a maddened bellow to my left and the animal was on top of me. I hurled myself sideways and a vicious sweep of the horns missed my head by inches and struck the muzzle of my rifle such a blow that it almost tore it from my grasp. The steer jumped over me with his front legs, carried forward by his own momentum. I was lying under his belly when he turned. As he did so he half dropped to one knee as if to kneel and gore me into the ground. It was a faltering step, his tongue and muzzle were soaked in blood, but he meant to kill me. I thrust the rifle into his face as he came and pulled the trigger. Brained, his limbs collapsed and he fell dead across my middle. It took the combined efforts of the two Indians to roll him clear. We had earned our meat the hard way.

Fifteen days had passed and we were within a dozen miles of the Berbice savannahs. Here we would hand the herd over to some negro cowboys who would march it from there to the village of Takama on the Berbice River. At Takama the steers would board a steamer bound for New Amsterdam at the river mouth. From there a railway took them on the final leg to Georgetown.

So, the journey for myself and the seven Wapishanas was nearly over; but that night it rained.

The deluge began shortly after midnight. It came down in torrents for about two hours, killing our fires, drenching our hammocks and supplies. We cut palm leaves along the banks of the rapidly rising creek that flowed through the camp, and huddled shivering under them. The rain eased at about 2.30 to a melancholy drizzle and we seized the opportunity to drive the herd across the flooded stream. Fortunately many of the animals had crossed voluntarily the evening before, searching for better pasture at the far end of the paddock. Encouraged by their presence on the opposite shore the other steers swam readily across.

Our problems however had only just begun. Ahead lay Yawikuri creek which was a fair-sized tributary of the Berbice. Normally at this time of year we could expect about waist-deep water and a good sandy bottom, no problem for the cattle. Now with at least 3 inches of rain on the ground it was everyone's opinion that we would meet a sizeable flood.

Shortly before noon, while still a mile or two from the Yawikuri, we were met by a wet and bedraggled negro riding a mule. It was impossible to stop the herd in the middle of the trail. If kept moving they tended to follow the line of least resistance but, held up, immediately became uneasy and stampeded into the surrounding forest. To prevent this I kept the drive underway and the negro dismounted and fell in step beside me.

"Baas," he said gravely, "water really big."

"Can we cross?" I replied anxiously.

"I ain't think so," he answered. "She runnin' like mad, an' the bridge cover."

"But you got across," I challenged.

"But this is a jackass," he replied waving his hand at the mule. "An' cow ain't got sense like jackass."

"You and I had better run on ahead and see," I told him. "My boys will hold back the cattle and walk them as slowly as they can to give us time."

The negro mounted his mule and set off, with me jogging along at his side.

Yawikuri creek was not the welcome sight it should have been. Twenty-five days had passed since we left Dadanawa and the end of the trail was less than 200 yards ahead. Before me lay a surging, mean, menacing sea, plundering the trees for 50 yards on each side of its normal banks. Great heaps of driftwood and giant tumbling logs heaved and rolled in the current. There was no sign of the bridge, for the level was at least a dozen feet above normal. My immediate reaction was to turn back; but even that was not easy. It would mean stopping the herd and forcing them to about turn in the narrow confines of the trail and go in the opposite direction. There was a strong possibility that such a manœuvre could cause a stampede. I looked across the 120 yards of turbulent water. I could see the long green ungrazed grass of the Berbice savannahs through the trees and the negro cowboys standing with their ponies. Maybe the herd would smell the grass and the open plains that for fifteen days had been denied them. Perhaps they would see that the gateway from their jungle prison was almost within their stride. All they had to do was swim, and swim straight. We would try it.

I stripped the saddle off the mule and sat astride him with the water swirling around his shoulders. We were some 50 feet into the flood and I planned, as the lead steers first appeared in the trail, to head for the sunken bridge. Hopefully the herd, seeing an animal in the river ahead of them, would decide to follow.

Meanwhile my negro companion had doubled back and rejoined the Indians. Their job was to force the herd into a run when they were within 150 yards of the crossing. If luck was on our side they would hit the water at a gallop, and be swimming before they had time to change their minds.

There was not long to wait. Soon the boys were shouting and hooting to wake the dead. The rumble of hooves drew nearer.

I drove my bare heels hard into the mule's ribs and moved forward. The water got a little deeper at first and I could feel the ground soft underfoot. It would be pure luck if I should locate the bridge. The negro told me it was only 6 feet wide and any moment I expected to step off the submerged bank into deep water.

Suddenly the herd broke into view at a dead run and I rode the mule for all I was worth. The first steers hit the water in a solid wall of rattling horns and jostling muscle. They saw me, and were after me.

Ten more feet and I felt the bridge. The mule kicked it, stumbled, and then found a footing. It was only belly deep; if only I could take it slowly we would walk right across. But even as the plucky animal took its first cautious steps my back was drenched in spray and the herd was upon us.

In seconds I was surrounded by a solid mass of steers and swept before their angry tide. I threw a glance over my shoulder and saw a pair of ugly horns set above ferocious eyes aiming for my back. Frantically, I threw myself sideways from the mule. There was another steer, red and white, its back covered in bleeding scaly sunburned scabs right alongside me. I fell across his withers and

(*top*) A tamandua carrying its young

(*bottom*) A giant ant bear carrying her cub saddled on her back

(*left*) The very rare pygmy anteater

(*right*) A capybara, the world's largest rodent

A sting ray speared in the New River

(*above*) Two giant otters inspect a basking caiman

(*below*) Piranha

grabbed at his horns. He lunged forward and went over the bridge; suddenly we were swimming.

His back sunk below the surface and I straddled him like a horse, holding his horns like a ridiculous pair of handlebars. All at once, there were animals everywhere, swimming for their lives, and all inevitably being dragged downstream by the fierce current. Anxiously, I looked back to shore. Longhorns still on the bank were fighting their way to get into the river. Now they were guided by the leaders, no longer on the bridge, but drifting downstream in a line of bobbing horns; following the least line of resistance, the centre of the flood to certain disaster. Then José and the Indians, and the negro boy, were there by the shore. They needed no word from me, and bravely plunged in, striking out strongly for mid-stream to head off the leaders.

To my left I saw the unfortunate mule, completely sandwiched in a *mêlée* of cattle. His eyes were wild with fear, and the bridle was entangled in a set of horns. As the steers compressed him betwen them he fought desperately, struggling upwards until he rode their backs with his foreparts. I had to get across to him; to cut free the bridle and disperse the steers before they forced him under.

Animals by this time were packed so tightly together that they formed a bridge. Already they had begun to mill; a deadly process of circling into an ever-tightening mass, until the ones in the centre are pushed under. I reached out and grabbed the bony spine of the nearest Longhorn and slid off the back of the one I was riding. Cautiously I scrambled over him on hands and knees to the next one. So I went, like a crab, across the backs of the heaving herd to the mule. As I reached out for the bridle the steer who wore it tangled in his horns lurched away from me. The reins tightened, jerking the mule's head and throwing him into a frenzy as he felt himself stretched by the neck while his pelvis remained in a vice-like grip between the other animals. Hurriedly I hacked at the reins with my knife. They parted, and a moment later we swung into the trees.

We hit about 150 yards below the bridge. The river was flooded right out to the edge of the open savannah which I could see a scant 50 yards away through the forest. But the water was still deep, and the steers immediately began to climb up on each others backs as they piled against the trees. By this time José and the vaqueros had reached me, and together we battled to free the herd and head it for shore. My first concern was for the mule who was in imminent danger of being drowned or gored. Fortunately, as the *mêlée* broached against the trees the sandwich that held him broke up. I swam towards him, grabbed the bridle and struck out with my free hand for the shore. All about me animals were finding their own path through the trees, and suddenly we were on solid ground.

I left the mule standing shoulder deep in the shallows and swam back to help the boys. There were steers tangled by the horns and legs in sunken lianas, and there were steers with their heads wedged in the crotches of trees. Sadly, there were some already too late to save, floating lifelessly with only their bloated bellies above the surface. It must have taken two hours to free the last survivor before we staggered ashore and collapsed, exhausted, in the warm green grass. Later we counted the herd, 438; nineteen had drowned.

For those of us who had come from Dadanawa it was the end of the drive, only the long and weary road home remained. The black cowboys would be in charge from now on, and the Longhorns were still a good distance from their destination. They would hold them for several weeks at Tacama, herding them loosely on the tall grassy Berbice plains. The steamer that transported them to the mouth of the river only came once a week and would load fifty or sixty at a time. Then it would chug overnight to the port of New Amsterdam, where the cattle would be off-loaded into corrals, given water, and then put on freight cars for the 70-mile journey by rail to Georgetown. They would be slaughtered immediately on arrival, and the local butchers would buy the beef as it hung in the abattoir. There were no cold-storage facilities in the city; but a large consumer population, and a perennial shortage of meat from other markets, ensured prompt disposal. I never visited that phase of the operation. The negro cowboys were employed on a full-time basis by the ranch, but they had their own foreman and received their wages direct from the Georgetown office, via the same steamer that ferried the cattle. The economics of raising beef points to only one objective, putting the finished product in the butcher's shop. But to a Dadanawa vaquero it seemed as if that aspect of the business didn't exist. The Longhorn was a wild animal that we hunted and chased, drove into corrals, lassoed, branded, herded for miles through desolate country, swam across rivers, pitted our wits and horses against in the fury of a stampede. He was a magnificent challenge, not an animal to end his days on a gory meat hook. I think all of us secretly hoped, when we handed our herd over to the black cowboys at Yawikuri creek, that somewhere between there and the end of the trail they might all escape and find their way back to the Dadanawa.

The march back to Surama took a little less than half what it had taken with the cattle. On our way, we passed the other herd. They had sustained no losses, and without further rain, would doubtless meet the Yawikuri quietly flowing beneath the bridge.

While we walked I thought of the horses, rested and strong. Five days hard riding from Surama and we would be back among the familiar pastures and hills at Dadanawa. It would be good after twenty-one days on foot to get back in the saddle.

We arrived at Surama early in the day, and I was anxious to mount up and put another 20 miles behind us before dark. I greeted the two horse boys and immediately enquired about the health of Cang and Butterball.

"Cang is all right," he told me nonchalantly.

"But Butterball is probably dead."

"Dead!" the words hit me like an electric shock.

"What do you mean, dead!" I said and shook him by the shoulder.

"Cabra bundo!" he replied hastily, the very tone of his voice insistent that it was nothing he could be responsible for.

Cabra bundo was the Brazilian name for the dreaded horse disease trypanosamiasis. Tryps, as it was commonly called, appeared in irregular outbreaks. A fly-born ailment which attacks the blood stream, it causes an invasion of the red blood corpuscles by their white counterpart. At that time there was no cure for it, and the disease, once the symptoms showed, was 90 per cent fatal.

The apparent lack of concern on the Indian's behalf for the life of my horse was not something he could be blamed for personally. The Amer-Indian is an unemotional creature, and the life or death of animals, and even his fellow men, is something he takes for granted with no obvious feeling.

I found Butterball down by the creek. He was terribly thin, and shaking in the hindquarters. There was a large pudgy oedema which extended along the underside of his belly, and his head was swollen. Both certain signs of the abominable *cabra bundo*.

He seemed to recognize me as I approached and whinnied weakly. I led him a few steps into the shallow water and he bent his head to drink. The tired pony sucked as if the water did not want to flow and the throat struggled to swallow it. He raised his head and the water ran back through his clenched teeth. I led him slowly back onto the bank.

Butterball looked at me with the same large brown eyes that had laughed when his lightning turn behind a renegade cow had been too much for a greenhorn cowboy. But they were tired now, tired of the agony of biting flies and the penetrating icy stares of waiting vultures. He could still hear the thundering hooves of the herd, the yell of the vaqueros, and the smell of the dust from a thousand round-ups. Given a chance he would have run his last gasp behind a runaway calf. Sorrowfully he hung his weary head and tugged weakly at the fresh green shoots of grass that peeped through the sand. If only he didn't have to die so far from the friendly familiar pastures of the Dadanawa he loved so well.

I cocked the Smith and Wesson and levelled it carefully between his eyes. It was a sad way to say good-bye to a friend.

X

BRAVE HOUNDS AND SHIWISHI-WIN

Life sped quickly by on the ranch. I never found it lonely, despite the fact that during my entire life there Jimmy Brown was the only companion other than the Indians. Even as time rolled by, Jimmy insisted on keeping to himself and we rarely exchanged more than a few words a month. I found, in fact, that the solitude suited me, and, except on round-ups when a lot of men were essential, I preferred to ride the range alone. This also allowed me to study the wild life of the country more closely. Deer watched as I rode by, birds followed me from tree to tree; often a falcon would hover overhead, waiting for my pony to flush an unsuspecting quail from the grass ahead. When I rode with the vaqueros, twelve of us, clattering through the brush laughing and joking as cowboys do anywhere, terrified every animal and bird within a 5-mile radius.

Years passed like months, and little seemed to change at Dadanawa. Yet, perhaps this was not so remarkable. We were completely isolated from the outside world. Visitors amounted to only two or three per year, usually government officials on inspection tours of the interior. Longhorns roamed the range as they had since the beginning of the century. They were just as wild now as they had been then. Round-up schedules were essentially the same. The first began in September and they continued almost without interruption until May. June, July and August were submerged under a barrage of 40 inches of rain. Life was a constant adventure: stampeding herds, wild broncs, ornery bulls, floods, swimming horses, swimming cattle and the ever recurring menace of marauding cats—mountain lion and jaguar. Then, in between, disease. Malaria, that left the body limp and weak from ague in a sweat-soaked hammock. Ten days of delirium during which one ate nothing, but alternately shivered then fried, occasionally falling into a shallow, troubled sleep. The scourge of the tropics, it has killed more men than all the battles in history.

We saw years of plenty; and years of drought, with blackened soils scorched by grass fires, and slowly dying cattle, hopelessly sunk in the mud of dwindling waterholes. Cassava crops were lost, and yet somehow, as he had for centuries, the determined Indian lived on.

Still we lived in hope for the best. When the weather was dry, and the savannahs were fanned by a cool breeze after the December showers, it was a beautiful country of sparkling clear streams teeming with fish. This was the time of year when the fortunate cow and mare would produce her calf and foal. It would be four months before the onset of the harsh rains and by the time they arrived the young ones would be big and strong. Some, however, would not be so lucky, for under the open-range system the breeding season was regulated only by natural instincts. Many were destined for birth in a foot of floodwater; while others might only escape to a lingering death as the bloodied end of their umbilical cord was infested by legions of blow fly maggots eating them mercilessly alive.

It was a strange phenomenon that few species of mammals arranged their family planning so as to produce their offspring in the comparative safety of the dry weather. Among the few exceptions were the giant river otters, who gave birth in January; the armadillos in April and May; the bizarre giant ant bear who could often be found with a myopic youngster clinging to her back shortly after New Year; and the grey savannah fox who invariably hid her cubs in a disused armadillo hole some time in March. For the rest of the denizens climate seemed to be a minor consideration. Yet, unlike the cattle, the strong urge for survival under the harshest conditions of floodwater usually saw them through. The jaguar and puma were a good example. They bred only once in two or three years, a litter of one to four cubs, after a gestation of 90 to 112 days. Birth invariably took place in the dry seclusion of a cave or hollow tree. It would be ten to fourteen days before the cub's eyes were opened, and two months before they would first leave the den on a hunting excursion with mother. Throughout this period, until they left her to live alone at eighteen months, they were guarded jealously.

If the mammals were erratic, the birds and reptiles were unquestionably calendar conscious when it came to breeding programmes.

The fish spawned while battling the rising floodwaters in May; in August the dragon-like iguana scooped out a nest in the hot sand and laid its eggs; while the water turtles did likewise between January and March, when the sand bars were fully exposed. The birds of prey, from the mighty harpy eagle to the tiny-hawk, nested between September and October; and the water fowl, teal, tree-ducks and muscovy only in May and June.

So in a year clearly divided into only two seasons, a month rarely passed without some significant event in the natural history almanac. A myriad of baby spiders on a web in September, a struggling mass of baby scorpions clinging tenaciously to mother's back in August, all were a demonstration of Nature's uncanny direction.

During these early years what had begun as casual observation on the Rupununi fauna gradually led to more intense studies. In 1959, the Royal Ontario Museum in Toronto sent a field-study team to the northern savannahs on a mammal collecting expedition. I met the leader, Dr. Randolph Peterson, when I accidentally encountered his camp during a cattle drive. He had already collected more than 100 different species and was on the point of moving to another area. I invited him to bring his team to Dadanawa, which he subsequently did, and collected a similar number of specimens along the southern edge of the Kanakus.

Dr. Peterson was primarily interested in the small mammals, bats especially. This was a particular education to me as I had never really taken much notice of bats, although there were plenty of them flying around, mainly vampires. In fact, the latter were responsible for spreading paralytic rabies amongst the cattle and the herd sustained varied losses from the disease every year.

Despite the unsavoury habits of the vampire, it is unfortunate that the bat family as a whole has never been one that endeared itself to the general public. From ancient times they have been associated with witchcraft, myths and various superstitions, inseparable from their cousins, the vampires, and supposedly equally devoted to the extraction of blood from hapless mortals. It is true that all bats have sharp, well-developed teeth, but then so do we and so do horses. Only the vampire itself will intentionally attack a human being. What is more, out of the incredible variety of Chiroptera known to inhabit the Guianas, there are only two species of vampire, *Desmodus rotundus* and *Diaemus youngi*. Neither of these are much more than 3½ inches in body length, and their wing spans stretch to only about a foot. Furthermore, both have a decided preference for the blood of all other creatures before they will prey on man; and so horses, cattle, and even birds such as domestic fowl, are the most likely ones to be attacked. They are incredibly energetic hunters, and an individual's nightly flight plan is often as much as 50 miles. The Rupununi savannahs are infested with vampire bats, many of them infected with rabies. In a bad year, the Dadanawa herd would sustain losses in excess of 3,000 head, and yet there has never been a case in that area of a human being dying of this dreadful disease. In fact, I only heard of one person in the cattle country who was ever bitten. However, in forest regions where livestock was not available and wild game scarce, a few people were occasionally attacked. Of course, not all vampires are infected with rabies, but they live on an exclusive diet of blood so they must hunt for it regardless. If they are not infected, the bite is completely harmless—if they are, then without vaccination before the onset of the symptoms, the consequences are 100 per cent fatal.

On one collecting excursion with Dr. Peterson we discovered a roost of *Desmodus* in a huge hollow tree, most of which we were able to capture by erecting fine mist nets across the opening. We simply frightened the sleeping bats who then flew into the nets and became entangled. At the end of the day we'd caught more than 300. In later laboratory tests all of them proved to be rabid!

Many times I have spent a night outdoors sleeping in my hammock slung between trees, and a dog lying right under it would be bitten so many times before morning that he would be covered in his own blood. Yet, always, I or a companion would be completely untouched. The vampire's bite is inflicted with an extremely sharp pair of chisel-like incisors in the upper jaw, and the blood is lapped, not sucked. There is a V-shaped groove in the animal's lower lip to facilitate this, and an anti-coagulant in the saliva which ensures a constant flow. It pulls itself across the victim's body by means of hook-like thumbs on the leading edge of its wings. The whole operation is so delicately done that the victim is rarely awakened.

Well, enough of vampires. They may be the most notorious but they're probably the least interesting of all the Chiroptera.

There are bats that live on insects, and others exclusively on fruit. While one of the larger varieties, *Vampyrum spectrum* (not a vampire despite the name), preys on other bats, lizards and even small birds. Yet with a 30-inch wingspan it is still a far smaller animal than those known by myth and fiction. Slightly smaller than *Vampyrum*, and probably the most interesting species of all, is *Noctilio leporinus*. This bat lives on fish which it catches alive, at night, along the surface of the water. At one time it was thought to occur only in Suriname, but Peterson proved it to be widely distributed in Guyana and Brazil. Since then it has also been found on the island of Trinidad. We were able to catch a number of specimens by stretching our mist nets across narrow river banks a foot or so above the water. It is more commonly known as the bulldog bat because of the pouch-like formations of loose skin in its cheeks and lips. Other unusual features are a beautiful rich red body colouring, long slender upright ears, and powerful talons on the feet. Colonies of up to seventy in number roost in rock crevices and hollow trees, but at dusk when they leave to go hunting they split up individually. The technique used to catch the fish, which are only small minnows, an inch or two in length, is quite simple. The bat merely flies very low along the surface, trails its talons in the water, and then clutches its prey on contact. Often the fish is eaten in flight and then the bat resumes hunting immediately. However, the bat is not always the hunter, as Dr. Peterson and I witnessed one evening when a large splash marked the end of a low-flying *Noctilio* in the jaws of an equally voracious rainbow bass.

Bats have very poor eyesight and yet they are able to fly swiftly amongst a maze of trees and branches in the dead of night without a collision. They do this by means of a unique built-in radar system. While flying each bat sends out a series of high-frequency shrieks, too high pitched for man to normally detect. These bounce back from objects in their path—not only from obstacles to be avoided, but also from potential prey, such as an airborne insect. The important difference between bat radar and the kind we are using today is that ours will only work in the atmosphere, while the bat's will operate under water. Obviously the fish-eating bat is equipped with something additional that allows him to detect a swimming minnow; and if we can ever learn what this is, it may bring to light a new method for detecting shoals in the commercial fish industry, or of course, submarines. The question becomes even more complex when we ask ourselves how does the bat's radar interpret the difference between echoes received from another bat that may be crossing ahead of it, and that of a branch to be avoided, or an insect or a fish to be caught? Even more complicated, how does it recognize its own 'beeps' from those of a hundred other bats that it may be flying with in crowded formation? Now add to these unanswered questions one that seriously effects the lives of thousands of cattle and people in Latin America. How can a vampire bat carry one of the deadliest viruses known to science, and at the same time be totally unaffected by it? The little appreciated and much maligned family of Chiroptera certainly warrant a great deal more attention than we have been giving them.

My introduction to these unusual creatures generated within me a great enthusiasm to become a collector. There are no less than 1,300 different species of bats classified from throughout the world. In the inaccessible forest regions of the Amazon there are sure to be many yet to be discovered. They have lived on this earth for over 60 million years, one of our planet's oldest remaining inhabitants. They have given me a broader base for my studies of the wonderful world of natural history.

Thus, seven years went by like the passing of but a few moons. Perhaps all that changed was within myself. I grew older, mastered the language of the Indians and learned their ways. At times I would go for six months without speaking a word of English, even dreaming at night in Wapishana. Yet it was far from a hermit's existence, and by 1960 I had built my own circle of friends.

They lived some 200 yards from my house, in a large roomy compound with thatched roof on the top of a hill. The head of the family was a fine fellow named Mr. Manager, from Hertfordshire, England. His favourite wives were Marie, Bunny and Poppy. Marie and Bunny came from Sussex, but Poppy was from the Wapishana village of Shea, 40 miles to our east. Mr. Manager was all for integration.

There were twenty-six in the family altogether. They lived more dangerously than anyone else in the territory. They were foxhounds, and their job was to track down the feline rustlers that plundered the range. Pledged to eschew all bar the smell of cat, these dogs would die before retreat when the jungle king turned to fight. Yet jaguar and mountain lion were not the only foe they faced. Lurking in the pools of quiet meandering creeks were caiman, anaconda and boa. While beneath a fallen tree or shaded rock, ill luck could chance upon the deadly coils of bushmaster and fer de lance. A jaguar could break a dog's spine with a single blow of a forepaw. His name is taken from the Indian *yaguara*, which means, 'he who kills with one spring'. *El tigre* is the only big cat who slays his prey via direct injury to the brain. His attack is always at the head, seeking to drive his fangs through the top of the skull. Rarely is he defeated in this aim.

There were many jaguar and puma in the forests surrounding Dadanawa, but only those who took to destroying livestock were hunted. Some were tracked down within a few weeks of beginning their rampages, for seldom after tasting the ease with which domestic animals could be slain would they return to their normal wild prey. Others displayed an uncanny cunning and waged an incredible toll of destruction before they were brought to book.

The cattle slayers of Shiwishi-Win fell into the latter category.

What Shiwishi-Win meant, even the oldest Wapishana in the savannahs couldn't tell me. It was a name handed down from extinct tribes, once warlike Aterods and Tarumas, who died along with their languages sixty years before.

Carved by nature from boulder-strewn slopes of the Kanaku Mountains, Shiwishi-Win stood like a giant dome of black granite, its easterly face weathered ghostly grey by an age of prevailing wind and rain. The base of her towering cliffs was guarded by dense thicket laced with flesh-searing razor grass and spiny-palm, every limb and leaf armed in needle-sized thorns. This inhospitable fortress overlooked a remote section of the ranch. Within the caves and crevices of her walls lurked some of the most elusive predators that ever prowled the range.

It was September 1960. The seasonal rains had just ended. Turbulent, swollen rivers were content once more to laze along within their banks. Cattle, earlier driven from the foothills of the Kanaku's by the influx of biting flies, planned a return to the lush pastures of the forest edge. But green grass was not all that lay in wait.

I was debilitated by a severe bout of malaria which had only subsided but a few days. For a week my stomach had refused all bar water and coffee and I had lost nearly 10 pounds in weight.

Except for a few short rides near the ranch house, I spent most of my time sweating out the last remnants of the fever in a hammock slung in the cool breeze of the veranda.

Unfortunately, this period of quiet convalescence was soon disturbed by the arrival of one of the outstation foremen padding up my front stairway in his bare feet.

"*Baidecour, baas* [a big cat, boss]!" said Caitano grimly.

I lifted my legs over the side of the hammock and sat up. "*Tikky baidecour druyan tapi-err Shiwishi-Win et, yatum vaca, pommytonky bizerro,*" he announced.

This was indeed bad news: two cows and four calves killed by cats in the past few days.

"*Kunam kader baidecour?*" I asked him, for the word *baidecour* was Wapishana for the feline family as a whole. Were we dealing with jaguar or puma?

"*Auna waichpan,*" he replied uncertainly. "*Mara elip d'yan!*"

Whatever they were at least we knew there were more than one.

"*Waikanan wom macun,*" I told him. Hopefully by the morning I should feel a little stronger, for the granite cliffs of Shiwishi-Win were no place for a sick man.

The Land-Rover and trailer brought us to within 3 miles of the mountain. To camp nearer would only risk to alert the quarry of our presence. Hunting both jaguar and puma calls for patience. Not for them the trap or prepared bait; or to wander abroad by day. I had a standard procedure when searching for livestock killers. Early the following morning it was under way.

All hounds remained in camp on the leash, resting under shade, for by eight o'clock there would already be a fierce sun. Meanwhile, four of us on horseback spread out, covering every likely haunt that might produce a fresh spoor or, better still, a recent kill. I took the shortest route to ensure my return to camp before the others—to be ready to mobilize the hounds without delay should anyone strike a hot sign.

Spoor we found plenty of, though all cold. It had not rained for several days and the sand bars in dried-out gullies had been richly autographed by feline passers by. A number of circling vultures over a thicket some distance from the mountain foot caught my eye. I rode over to investigate.

Approaching from downward I recognized the smell of putrefied meat. On investigation I found an animal about three days dead. It was a yearling heifer with horns only a few inches in length. What the cat had left the carrion crows had disposed of, and skin and bones were strewn over a wide area. The ground was hard but a piece of sand nearby revealed the attacker's spoor. Its prints were of medium size, 11 inches in circumference. I guessed the owner to be a puma, probably a male, for the females are considerably smaller. Closer inspection of the carcass confirmed this, as there were no marks on the skull other than the gnawed ends of the jaw bones. Pumas kill like leopard, lion, cheetah and tiger, with a fang hold in the throat, the prime cause of death being asphyxiation. There was no evidence that the culprit had returned to the scene during the past twenty-four hours. In country so rugged it was pointless trying to follow him with the hounds. Give him time and he would strike again.

Two uneventful days passed by, and the morning of the third looked like being another. I had just returned to camp and pulled the saddle off Cang when Caitano galloped up with word of a freshly-killed domestic pig 2 miles away in a wood along the river bank. Fortunately, the other boys arrived a few minutes later, and without delay we loaded Mr. Manager and three other hounds into car and trailer. Soon we were testing the limit of its springs over the rough, rocky ground before us.

The kill had been dragged into a dense bramble thicket, only betrayed by two black and white king vultures who flew off at our approach. The carcass had been well concealed under a pile of dry leaves carefully arranged by the cat. I brushed them away and found the dead pig only partly eaten. The stomach was ripped open, but the entrails were still intact and only the ends of the ribs had been devoured. Fang marks on the throat confirmed that it was the work of a puma. It had only been killed an hour or two earlier, the body was still warm.

We crawled around in the thicket seeking to reconstruct how the attack had been made. This did not take long, for we soon found an obvious, well-blooded, beaten-down tunnel through the cover where the carcass had been dragged. Following it for twenty paces brought us to the dried-out bed of a creek where churned up sand and puma tracks clearly showed where the kill had taken place. Apparently the pig, a stray from the ranch, had been ambushed while searching for water. The stage for the hounds was now set.

As soon as Manager got a scent of cat around the dead pig, the hair on his back stood up like a brush. I slipped his collar and urged him to go. After circling the spot three or four times, he started off through the woods with a series of bugle-note barks. With the other dogs still on the leash we plunged into the undergrowth in pursuit.

By now Mr. Manager was some way ahead, yet moving slowly, carefully sorting out the trail. This was not the time to release the other dogs in the hope of speeding up the chase. They were more likely to confuse and perhaps even obliterate the scent, swarming furiously about the jungle, rather than assist. The Manager needed no help at this stage. When the cat was up and running then the others could be freed, to overtake their tiring leader and press the quarry until bayed.

Suddenly Manager, 50 yards ahead and obscured by thick cover, gave several deep, throaty, booming barks. Immediately he began moving away, bawling with every step.

At once I ordered the release of the other hounds in support. The thickly matted vegetation was no barrier to the solid wall of dog flesh that barged zealously through, while we assaulted it with our machetes and strove to keep at least within earshot. The chase moved directly away from us for about a quarter mile. Then, as it invariably does with a cat, began to circle back. We paused as the noise of the pack came closer and closer. They passed a stone's throw to our right, yet none of us saw the puma which must have run almost through our ranks. For a minute, the tongueing ceased and I feared the hounds had lost the trail; but soon it started again, even more ardently. The hunted had only dodged them momentarily. Moments later it treed.

We found the pack barking in a frenzy at the foot of a mere sapling. A fine male mountain lion balanced precariously, all four of its feet in one spot, on a narrow fork barely 10 feet above the ground. Powerful shoulder muscles rippled in readiness, and 120 pounds of angry cat spat viciously down at the leaping dogs. It saw us at once and began shifting position uneasily.

"*Kademen peu baion, rikaitan-na* [quick, shoot, he's going to spring]!" There was justified urgency in Caitano's advice. The cat was about to spring and a certain first shot was essential. Any mistakes would mean the probable death of one or two brave hounds.

The angle was such that to brain it required a shot under the right eye. Quite a simple one as the range was barely a dozen feet. As I fired the animal leapt into the air. A second more would have been too late. The impact of the bullet spun it sideways to crash into the middle of the pack. They were on it at once, but there was no resistance.

The first of the stock killers had been brought to book.

I awoke next morning shivering from a recurrence of malaria. My hammock and blanket were sodden with perspiration, and a pre-dawn breeze on my damp body had chilled me to the marrow. I got up and walked over to the fire. Caitano was sitting by it, and remarked as I approached, "*Kary-nyan waichpan* [you look sick]!" My knees felt weak, and showed it.

I let a cup of strong, black Brazilian coffee trickle down my throat. Across the savannah I could see the bleak uninviting walls of Shiwishi-Win, shrouded in an early morning mist. By eight o'clock I should feel either a little better, or dead. I decided to remain in camp and send out the boys alone to check for spoor.

Three hours later two of the Wapishanas returned with nothing to report. My condition had not improved and trembling alternated with periods of icy sweat. Ten minutes passed before Caitano galloped up. His horse heaved heavily at the flanks and foamed at the bit. It was a call to arms.

He had encountered a travelling Indian who told him of two jaguars that he had seen last evening in the very act of attacking and killing a calf. The victim's mother made a brave attempt to drive them off, bellowing and charging, hooking with her horns. Immediately the smaller jaguar, presumably a female, abandoned the dead calf and sprang at the cow, both meeting in a head-on collision. Fortunately the Longhorn's charge was more determined than the cat expected and the deadly fang hold between the horns was missed. The cow, now terrified for her own safety, swung around, dislodged the jaguar, and ran, making good her escape. Neither cat made an attempt to follow. This was not surprising for jaguars have little more than 70 yards of sprinting speed, and are no match in a foot race with a fleet-footed Longhorn. Instead, satisfied with their kill they proceeded to drag the calf between a rubble of boulders at the mountain foot, within the long afternoon shadows of the towering rock called Shiwishi-Win.

The Indian had taken Caitano to the scene but they found nothing save bloodstains, spoor, and trampled grass. There was no trace of the kill. Presumably the calf had been carried to the jaguar's lair.

After listening to the story I found the three Wapishanas watching me expectantly. The hounds were already on their feet, whining excitedly, straining against their leads. They knew there was something in the air. Malaria or not, I had no choice. I drained the last drop of coffee from the jug, then picked up the rifle.

The Land-Rover bounced, battered, groaned, yet carried us to within one mile of the mountain. From there we would proceed on foot, the three boys leading six hounds while I followed shakily, carrying nothing but the gun.

A creek flowed between us and the cliffs which we forded knee deep. As we climbed down the bank a tapir stampeded from nearby cover and plunged with an immense splash into deep water. For several seconds it swam submerged, but then suddenly surfaced and caught a footing on the opposite shore. A mighty heave and it catapulted up the steep, slippery slope in a cloud of spray.

The mist had cleared and Shiwishi-Win posed statuesque against a blue sky. Somewhere within the tunnels and caverns of this stronghold lay two jaguars, alert, suspicious, scheming, razor-clawed and iron-fanged. They had chosen the site for their domain well. Now we, the intruders, prepared to lay siege.

Between the boulders where the two cats had made their retreat the ground was too hard to show spoor. Nevertheless dry blood was profusely scattered and clearly revealed the route along which the prey had been carried. Carried, as opposed to dragged, because a carcass the size of a calf, deer or peccary required to be moved any distance by the cat is lifted practically clear of the ground. The jaguar grasps the animal in its jaws by the nape of the neck, then raises it as high as possible. To do this with a large carcass it has to crane its own neck so far back that it is actually looking skywards. The prey's body is straddled by the cat's forelegs which as a result move in a pronounced bowed motion.

We turned Manager loose amongst the boulders. His hackles went up immediately like a cornered boar, and he moved cautiously sniffing each rock in turn. A hound knows the difference between jaguar and puma scent, and knows he can expect a different reception at the end of it. Packs sometimes run a puma in wild, frenzied, high-pitched yelps, but the jaguar always calls for deep, throaty bellows. Distinct 'music' that the hunter is quick to recognize.

Mr. Manager picked his way slowly and silently up the mountainside. The trail from the afternoon before was about sixteen hours old. Normally too cold to be worth following. However, we knew that the quarry, gorged on beef, could not be far ahead.

My legs felt weak and I soon found the going difficult. Then Manager opened up on the scent for the first time, his booming bugle-like call echoing through the cliffs above us. The pace increased and the terrain became steeper. A few yards ahead Manager disappeared into a tangled thicket. It was impossible to penetrate any other way but on hands and knees.

The trail was still cold, but alone Manager was in danger of being ambushed by one of the cats. As it was no longer practicable for the Indians to proceed further with the other dogs on leads I ordered them all released. Two of the older ones, Poppy and Marie, immediately picked up the scent, acknowledging it with coarse, gruff barks. The Indians fought their way after them on all fours, while I did my best to keep up at the rear. Although jaguars are excellent tree climbers they rarely seek this line of retreat in the face of hounds. Supremely courageous, if unable to outrun their pursuers they will stand and fight. Every second counts in getting there with the rifle in time to save the dogs.

All at once, 100 feet above us the pack opened up in unison. The chase was on. Huge, smooth-faced boulders, 10 and 20 feet high, loomed all around us. My only thought now was whether or not I could last the race. Already I was short of breath, my throat felt as if it had been scraped raw with a file, and I could actually hear my heart pounding against my chest. On hands and knees we burrowed through thorn and bramble, torn and bleeding, eyes burning with sweat. I slumped forward on my face, exhausted, and tried to be sick.

"*Kademen baar-ana airee* [quick, he's in the open]!" Caitano shouted.

I staggered forward and sure enough a few yards further on broke out of the thicket. There the mountain side was bare save for scattered trees, and heap after heap of towering rock.

The pack was nowhere in sight, but their tongues had reached a crescendo and I could hear the Indians encouraging them wildly. We were only a stone's throw from our goal, but had to fight for every inch. The chase was moving around the foot of the main Shiwishi-Win precipice, the hounds though close were hidden around a corner.

A moment later I stumbled over a dog. He was a yearling, a veteran of three puma hunts. This had been his first jaguar. One look at the bite through his spine and I knew it was his last.

Now the Indians were shouting frantically for me to hurry. Above the din of dog and man were the savage roars of a maddened cat at bay. I scrambled on as best I could; at last there were the men, 10 feet above me on a ledge of rock. But it seemed unclimbable.

I handed up the rifle to the two Indians who, flat on their bellies, leaned out over the ledge. Then I reached up and was grasped by strong outstretched hands. They heaved, I kicked furiously against the granite wall with my feet. A final effort and I was up.

We were on a small plateau, covered in sparse wiry grass and trees. Fifty feet away a jaguar was cornered, surrounded by the frenzied hounds. As I watched, it sprang at the nearest dog, but was immediately pressed from behind by the others, causing it to abandon its attack and swing around.

Suddenly Caitano thrust the rifle into my hands and I worked a round violently into the chamber. As I did the cat broke through the pack and disappeared amongst nearby rocks. We urged the dogs after it and followed at a run. But this stock killer had no intention of trying to elude the hounds. It was ready to fight to the last.

I dragged my malaria-wracked frame over the last giant boulder. Thirty feet away the jaguar had its back to a cliff and was swatting dogs like flies.

To save them I needed a brain shot. I staggered to my feet, flipped off the safety catch, and moved in close. At 10 feet I was looking over Manager's back. I couldn't see the sights for sweat and the jaguar was flashing its head from side to side. With vicious roars it made short rushes from one flank of hounds to the other. The all-important brain was a dangerously difficult target. Instead I decided to try for the spinal column in the neck. For a split second I saw the black bars across its broad spotted chest. I aimed slightly above and jerked the trigger. The bullet sent the cat rearing on its hind legs, and it fell amongst a mad *mêlée* of fighting hounds. I rushed forward, thrust the muzzle between the dogs against yellow fur, and fired again. The battle was over.

The animal was a large male, its gut bulging from a heavy meal. Where was its mate?

That could wait for another day, I thought, and threw down the rifle.

"*Ko-ygenna wom err-now* [Put the dogs on their leads]," I said to the Indians; and as they tied up the brave but tired blood-stained hounds, I nestled against a comfortable boulder. In a moment I was lost in an exhausted sleep.

XI

LEEMO AND CO

One might think that an annual toll amounting to hundreds of cattle, horses, sheep and pigs, slain by marauders of the feline kind, not to mention faithful hounds, would have led to an intense dislike of the cat family as a whole.

On the contrary, this could not be further from the truth. Instead I learned to love these animals that I was forced to hunt; to love their cunning, stealth, strength, speed and incredible ferocity. Whenever possible I avoided shooting the cattle killers, but trapped them alive, sending them to zoos, or relocating them in remote areas where there was no livestock to get them into mischief. It also became my ambition to tame and raise the big cats from cubhood. Being essentially nocturnal forest dwellers, they were virtually impossible to study at close hand in the wild. Apart from a few chance encounters, my experience with them was largely confined to examining their tracks and noting their behaviour when cornered by a pack of hounds. If I could raise them in my own backyard, allowing them to roam freely and naturally, I could maintain a schedule of serious daily observations. Without cages they would be virtually in their natural habitat, for the forest-lined river bank was but a few yards from my back doorstep. Animals with habits little known to man would suddenly reveal themselves.

By 1964 I had acquired not one cat but several, and the Dadanawa Ranch, in addition to being a cattle station, was gradually transformed into a cageless, wildlife observatory.

The animal that started this trend and commanded my deepest affection was a female mountain lion that I called Leemo. She was only a few weeks old when I got her, from an Indian who found her wandering about in the forest. He probably inadvertantly frightened the mother who perhaps had a second cub, and only had time to pick up one in her mouth and run off into the forest. Unfortunately, there was no possibility of ever returning the cub to its rightful owner so I assumed all responsibilities and became the foster parent. By some strange coincidence a similar incident produced a young male ocelot only a day or two later, which I named Beano. Despite the inherent mistrust between such species the pair soon settled down to a lasting friendship, and both became as tame as adopted alley cats. Unfortunately, natural instincts led them ultimately to mischief. Particularly Leemo, who soon became big and strong enough to pull down a Longhorn calf or an unsuspecting sheep. I went to great lengths to avoid such incidents but the speed and cunning of a wild cat produces tactics that cannot always be thwarted. There were times when some of the domestic stock mysteriously disappeared.

The mischief that surrounded her was further aggravated by the presence of two jaguars, Chico and Choo-Choo, who rapidly grew from cubhood to frightening proportions. They were male and female, and therefore reasonably compatible, but between them and Leemo stood the bitterest enmity. At times disaster loomed so close that even the entire human resources at Dadanawa were hard pressed to prevent it.

In the meantime while I struggled to avoid a clash between the major powers, the roll call of other animals in the yard expanded at an alarming rate. All at once I found myself the collecting centre for every waif and stray in the jungle. Never a week went by without some Indian arriving on my doorstep carrying a sack bulging with mammal or reptile. My undoing was a complete inability to say no.

Before long the house was alive with such things as boa constrictors in the rafters (working on their own programme of bat and rat control), two spider monkeys who had to keep a wary eye for prowling predators, and found the best way to do this was to remain out of reach on the radio aerial (which, unfortunately played the devil with our transmissions and reception); an awesome harpy eagle named Jezebel, with a propensity for attacking stray little Indian kids. Tiny, a 500-pound tapir who would rather blunder his way myopically through the fence than go round by the gate. A giant ant bear called Andy, who hid from Leemo in our overgrown vegetable patch. A margay cat, and a baby female ocelot named Domino, who hopefully was a prospective mate for

Beano. Jill, who was a paca living under the floorboards downstairs. A herd of deer in a fenced enclosure who had their own Indian bodyguard (as a visit from Leemo would have been disastrous). And, if this does not already convey a scene of utter chaos, there were four very rare, and very naughty wild dogs and an Australian dingo, who was a present acquired during a visit to the Georgetown Zoo. They ran together in a pack, and their specialty was tugging at the ant bear's tail, nipping Tiny in the hock, and ambushing both Jill and Domino. The tables however, were frequently turned when Leemo made an unexpected appearance and routed the four delinquents. In their haste to escape the red menace, they sometimes ran right into the waiting claws of Chico and Choo-Choo. The business end of a broom was on many occasions their only salvation. Without Irene's ability to wield it with unerring accuracy it would have been doubtful that they could ever have survived.

Actually, maintaining even a semblance of control amongst them would have been impossible had it not been for Irene. She loved them all and they adored her equally, although the cats had unusual ways of showing it and I had to constantly remind Irene to be careful. Without her, keeping any of the animals would have been out of the question. I spent so much time away on round-ups that Irene ended up looking after them even more than I did. This responsibility could never have been delegated to an Indian, as none of them worked for more than a few months at a time. The animals would no sooner have become accustomed to one handler than they would be faced with another. This could never lead to the kind of trust so essential in the safe handling of large cats. Most of the animals were raised on a bottle in the early stages. It was in this aspect of handling that Irene was particularly brilliant. She could take even the most difficult youngster and within a few patient hours have it sucking milk off the end of her finger and then taking the teat almost without realizing it. A few of the creatures, like Jill the paca, were so nervous that they only appeared at night. This was normal behaviour as they were nocturnal species, and the presence of cats in the yard made them understandably cautious. Jill came to the kitchen door every evening after dinner and Irene would sit down on a box and feed her anything from bread and milk to a coconut. Though Jill took food out of her hand, she would not allow herself to be touched and would never venture inside the door where she could have been trapped. She had an interesting habit of stamping her back foot, which appeared to be more of a mannerism than a sign of alarm. Pacas can run with great speed, but until we got Jill I never realized that they would break this gait and bounce stiffly up and down on all four legs like a rubber ball when frightened. This was particularly enlightening as a number of African antelopes, notably the springbok and Grant's gazelle, do exactly the same thing. Jill soon learned her name, and, so long as the sun had set, Irene had only to call once or twice before the paca would wriggle out from her den beneath the floorboards and come running.

It wasn't very long before the word got around the Wapishana tribe that I was building up a collection of animals. Being traditionally a hunting people they could naturally think of numerous species that were needed to make my menagerie complete. When they further learned that I was prepared to pay for animals that normally they by-passed as inedible, or hit over the head for amusement, their efforts were intensified. Rather regrettably, this led to the arrival at Dadanawa of a small percentage of rather undesirable creatures who would have been much happier, and infinitely more useful, loose in the jungle.

"*Kaena um nakan penoke'iree peu-at* [I've brought a most unusual animal for you]!" This was the standard approach made by all of the well-meaning would-be vendors. It meant they had brought me a *penoke-iree*—a beast in other words, but not an ordinary beast. A *penoke-iree* was the word for a very special kind of beast, one unknown except by myth, and perhaps now to be produced from the dark interior of sack or box for the first time. Inevitably, the mysterious denizen turned out to be a relatively common specimen, apprehended as it innocently crossed the trail only a few hours previously. Despite the accompanying sales pitch (which assured me that the animal was tame, particularly with children and if only there was one here now I would see), it was usually tied with a stout piece of cord or wire and ready to chew off the first set of fingers that dared go near it.

Whatever the species, I had no choice but to buy it. The Wapishana does not feel the same compassion for wild things that we do, which of course is a process of education. One has only to look back at some of the games our ancestors played with animals to prove this. If I didn't take it, the chances were good that they would kill it, rather than set it free, as soon as they left the ranch. Most of these creatures I acquired for a modest barter of fish hooks or coloured cloth, and then, with the happy salesman safely on his way home, released them along the wooded river bank behind the house. Admittedly, there were occasions when even a fish hook seemed expensive payment for some of these acquisitions. In one instance, after listening to the "*Kaena-um-nakan-penoke-iree*" piece, the contents of a grubby old sack were dumped at my feet. I barely jumped out of the way in time as an extremely annoyed 5-foot rattlesnake went slithering across my yard. I ran for a stick to pin it with, but before I could put my hands on something suitable it had escaped into a hole in the house foundations. On this occasion I made it abundantly clear, in the local vernacular, that instead of fish hooks, the owner could expect a kick in the ass. But when I turned to put words into action the mischievous Indian was already on his way.

My house soon began to prove rather inadequate for the constant stream of animals that paraded through it, particularly when one was a full-grown puma, shadowed by two rapidly growing jaguars and an ocelot. Unfortunately, the influx of fauna came only shortly after the building had been very carefully and expertly modernized by Jimmy. Not only had he made extensive repairs to the roof, but he had also installed running water, piped in from a large overhead tank drawn from a well by windmill power. After sinking a septic tank in the yard, Dadanawa proudly displayed its first indoor shower and toilet. Gone forever were those wild midnight dashes to the lonely little shack behind the garden, invariably postponed until the last painful minute, particularly during the wet season's torrential rain. No longer would I have to undress along the river bank in an evening cloud of kabouras, sand flies and mosquitos; or prod ahead of me with a stick in the sandy shallows to chase away the sleeping sting-rays. Gone forever also was the exasperating experience of seeing a brand-new bar of valuable soap, specially imported from Georgetown, slip through the fingers and sail away beneath the dark turbid waters of the Rupununi. In addition to the plumbing, Jimmy had also fixed the windows. New, plastic, long-wearing, non-corroding mosquito mesh, replaced the old, long-rusted wire variety. However, the rusty wire stuff had made a very effective animal barrier. Whenever it was necessary to restrict the

(*above*) A mussurana swallowing a rattlesnake

(*below*) The author struggles with the murderous coils of an anaconda in the Rupununi River

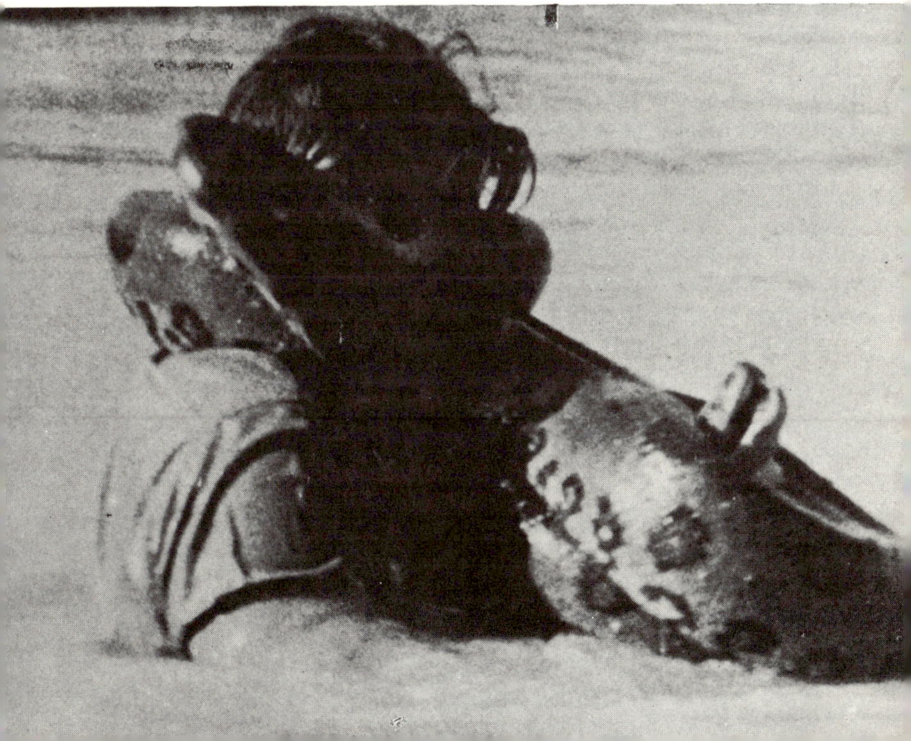

(*above*) The author and Jezebel

(*below*) Jezebel coming in for a landing (photograph by Ralph Nelson II)

(*right*) We transported Jezebel in the back of a Land-Rover. (*overleaf*) Jezebel at rest

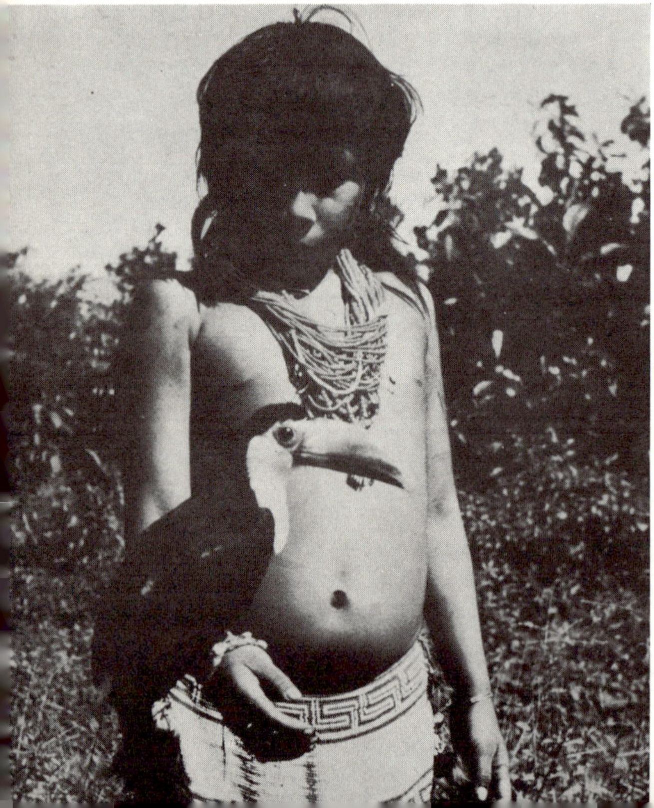

(*above*) A king vultur
(*left*) A Wai-Wai girl
with a toucan. (*right*)
The Wai-Wais give
Zero-Two-Zulu a car
ful inspection

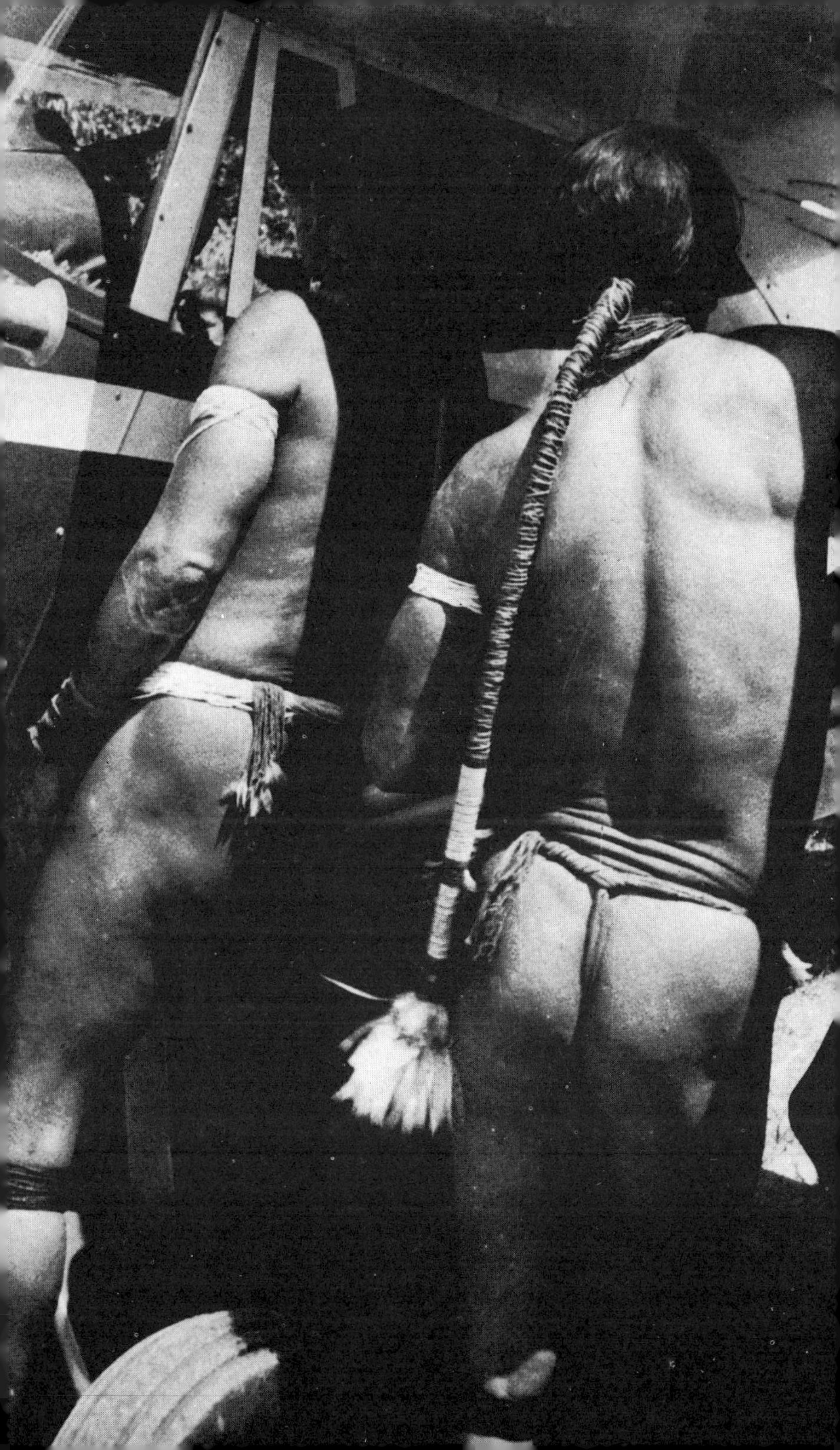

Shooting the rapids on the Essequibo

movements of certain creatures, I merely put them in the house and closed the door. Unfortunately, the new plastic mesh was designed only to prevent the passage of insects, not mountain lions or jaguars. Leemo shortly learned that if she stood on her hind legs, and leaned against it with her forepaws, the whole thing split down the middle. The look in Jimmy's eye as he saw the fruits of many hours of diligent labour systematically reduced to ribbons was ample warning that I had better do something about the behaviour of my animals.

So a special house was built for Leemo and Beano just behind mine, and one next to it for Chico and Choo-Choo. They were roomy, well ventilated, and the roof leaked even less than my own. When I was at home the doors were opened and all enjoyed complete freedom. They were fed in their house, and got so used to it that in the evening they were often already inside waiting for their dinner. Only at night did the doors remain closed so the cat's hunting activities were restricted. However, it soon transpired that this partial incarceration did little to prevent them from getting up to mischief.

On one occasion, after an absence of two or three days, I returned home and drove up in the 'Grey Ghost' to the front gate. I immediately called Leemo by name and Irene released her, together with Beano, from their house. They came pounding down the garden path towards me. Leemo several yards in the lead, her long legs moving with all the grace and agility common to her species, while Beano, tending to be somewhat lumbering and bow-legged, growled his annoyance at being left behind. I put down my kitbag and prepared myself for the inevitable impact. Leemo would always jump up and hit me in the shoulders with her forepaws, and if I didn't agree to fall down she would lock her forearms around one of my thighs and sink her teeth playfully, but sometimes deeply, into the flesh. Not only was this hard on the trousers, which were a scarce item in the Rupununi, but a set of teeth anywhere in the region of the upper thigh is a mite uncomfortable and better avoided. So the best course of action was to fall over backwards as soon as she hit, which I did; and then she was content to sit astride my chest and lick my cheek. Unfortunately, this display of affection was only slightly less painful than a bite in the groin, because a puma's tongue has a coating like the sharp edge of a carpenter's file. It works in exactly the same way, it rasps off the surface to get down to the meat. To add insult to injury the welcome home included the arrival of Beano, who to establish that I was once more in his territory, presented me with his rear end and sprayed me with urine. Such displays of affection not only convinced me that I was part of the family of animals, but also convinced the other *homo sapiens* that tried to share the Dadanawa hill-top that I smelled like one!

After recovering, and making my way with some impediment upstairs to my house, Irene arrived with a graphic account of the latest list of misdemeanours.

"Leemo tried to eat Marg'rit's pickney!" she started. Margaret was an Indian lady who cooked for the vaqueros in the bunkhouse. She was very fond of Indian men, which made her an excellent and steady employee, as one thing Dadanawa never had a shortage of was a bunkhouse full of Indians. The 'pickney', to translate from the Creole, was a child, in this case her son of about 7 years. Like most good boys of that age he had a grubby face and got up to mischief. His favourite prank was sneaking through the garden gate when Irene was busy and prodding the cats through the wire-mesh wall of their pen with a long pole.

"The next thing I 'eard was an 'orrible scream," Irene continued. "I ran out an' there was Leemo with one paw through an 'ole in the wire, holding 'im by 'is pants!"

"Good Lord!" I retorted, putting on my best look of grave concern.

"I ran up and beat 'er with the broom 'andle until she let go. The kid was bleedin' an' ran off 'ollerin' fer 'is mum," she concluded.

I sent word for Margaret to bring her wayward child for an examination. They arrived in due course, the mother totally unconcerned, her offspring, somewhat sheepish but apparently all in one piece. He was wearing a pair of dirty khaki trousers with the bottoms rolled up that would have fit any vanquero in the round-up gang. His mother turned him around to reveal the seat, recently patched with a large square of red and blue cloth.

"Sootan peu chooran," I ordered. He dropped his trousers to the floor.

His mother smiled, proudly pointing at her work of first aid. His bottom was covered in what appeared to be the entire contents of a tin of Band-aids. There was no swelling and he was well on the way to a complete recovery. He never went near the cats again, and from that day forward he had a new name—the vaqueros called him '*Shee-en dee-cap*', 'Scar-bottom'!

In addition to the installation of toilet facilities, Jimmy also put in electric light, powered for a few hours every evening by a small diesel engine. Naturally, to facilitate the passage of current between the various buildings it was necessary to set up a network of wires. Unfortunately, we had not reckoned on Dadanawa's two aerialists, Big Jacko and Little Jacko. The new wiring, as far as they were concerned, eliminated running the gauntlet of wild dog and jungle cat, when travelling overland between the buildings. Now they had their own completely personal commuting system. It wasn't long before Jimmy's neatly arranged parallel lines were suffering under the strain of two 30-pound gymnasts. First they drooped, and then they tangled. Then someone who hadn't noticed that they were entangled switched on the engine. The result was a Guy Fawkes firework display, a blown set of fuses, and a flow of invectives from Jimmy on the ancestry of the black spider monkey. When I heard him muttering words like 'shotgun' under his breath, I decided it was high time to construct a special primate residence. The design incorporated all of the things that monkeys find amusing: rubber tyres, the spokeless rim of a bicycle wheel hanging from the roof, empty cans that you can beat and make a noise with, and a network of wires that resembled the ones they had just finished wrecking. Unfortunately, neither Big nor Little Jacko seemed completely happy with their new surroundings. Try as we could to block all their escape routes, there was always some board that they could lift up or hole to wriggle through. They came and went pretty much as they pleased, sometimes with calamitous results.

I had barely swung out of the saddle one day after a routine ride when Irene came running to meet me, calling anxiously.

"Leemo, Leemo," she shrieked. "He tryin' to kill Jacko inside the house!"

For a moment I stopped in my tracks, my mind momentarily engulfed in a vision of complete and utter chaos. Then I ran as hard as I could, taking the veranda steps three and four at a time. The door at the head of the stairs was open and I charged in. My sudden intrusion froze both animals in suspended action. Leemo was 10 feet above the floor on the lowest rafter. She was flat on her belly with ears laid back and feet tucked well under her. Some 15 feet further away Big Jacko hung from another beam gripping it with only one hand and the tip of his shaggy prehensile tail. His face was a mask of terror, his free arm poised ready to propel him in whatever direction necessary to avoid Leemo's imminent spring. Around him the air was filled with frantic airborne bats and indignant buzzing hornets, rudely dislodged from their normally secluded perches by the violent invasion. Leemo's muscles flexed. I called her by name in an effort to distract her attention.

"Leemo, don't you dare," I said, then quickly, over my shoulder to Irene. "Bring me a big chunk of meat."

At that moment Jacko lost his nerve, swung up onto the beam and started along it. The movement prompted Leemo into action, but as she leaped the monkey sprang for a light bulb that hung from a long flex in the middle of the room. He climbed it rapidly towards the roof, a safe distance from the treacherous rafters that Leemo could use so well. If he would only stay there until I could get a hold of the cat; but the flex gave way, and bulb, shade, and Jacko came crashing to the floor. Leemo was after him in a flash. I jumped between them and took a wild swipe at her with my riding whip. The blow missed, but for an instant my body blocked her view. Jacko, with the commotion and shouting, was now as frightened of me as he was of Leemo. He jumped onto the table, hopped over the radio transmitter, and scrambled wildly up the aerial. Before I could breathe a sigh of relief Leemo aimed a blow at his rapidly moving rear end, missed, but scored a direct hit on the wireless and sent it skidding across the room. Jacko had reached the top of the wall that divided living-room from bedroom and dashed wildly along it. The puma clawed her way up the partition shredding in the process, a new and elaborate map of the ranch. Big Jacko's tail disappeared into the bedroom and the thump as he hit the floor was followed almost simultaneously with a heavier thud as Leemo landed after him. I burst through the bedroom door in time to see the terrified spider running on hind legs across my recently acquired bed, and spring desperately towards the gun rack. Leemo would have got him, but the bed linen caught in her claws and she fell to the floor frantically tearing and biting at the sheets. Guns, cartridges, belts, pistols and cleaning rods scattered everywhere as Jacko scrambled across the rack and leapt for another light bulb. He swung on the flex and reached the bookcase leaving the bulb to smash itself against a rafter. He paused for a moment indecisively. The puma answered with a fierce spring which came within a hair's breadth of catching the monkey's feet as he made up his mind and jumped once more to the top of the partition. Leemo clung tenaciously to the bookcase, kicking wildly against the glass doors with her hind feet, looking for a claw-hold to help her up. It swayed precariously under her weight, then toppled ignobly forwards to complete destruction. In an instant the chase was back in the living-room. Jacko stood poised on a table taking temporary refuge behind a row of bottles. Once again I threw myself between him and the cat, but to no avail. Monkey vacated the table and the cat landed amongst the bottles. Every brand of booze from the best scotch to the worst Brazilian paint-remover made instant cocktails on the wooden floor. Jacko went across my desk and headed for the open front door. The typewriter fell in Leemo's path and gave the monkey just the head start he needed. Outdoors in a straight run across the yard he wouldn't stand a chance. He had two choices when he reached the veranda stairs, run down them and be certainly caught, or leap for one of the electricity wires that ran to the corner of the building. It was a long jump, even for one so adept as Jacko. Twelve feet with his black spindly arms at full stretch and he was safe. He didn't stop swinging along the wire until he was high over the middle of the garden. Then he sat there, whimpering breathlessly, while Leemo stood beneath him on the lawn peering intently upwards. Irene arrived with the meat and I held it a few feet in front of her nose. She looked first at the meat then back again at Jacko, hoping perhaps that the wire might break. I moved away towards her pen calling her by name. She followed at a trot, stopping only once to look back at Jacko. I threw the meat into her house and a moment later she was under lock and key. Several minutes elapsed before Jacko dared come down from the wire, and when he did he ran as fast as he could for his own house. Safe inside, he sat on a shelf high up near the roof with his arms wrapped tightly around Little Jacko, and indeed it was a very long time before he thought about escaping again.

During the heat of the day Leemo loved to lie out on the veranda wall in the cool breeze, with Beano never far away. There she watched the world go by and waited for the cattle, sheep and horses to graze up close enough to the garden fence to be worth chasing. If I sat there with her she would sidle up alongside me and rub her sleek red sides against my legs, and purr like an overgrown house cat when I scratched her ear. She loved the taste of salt on my bare arms, and licked them with her coarse file-sharp tongue. Within just a few moments the skin would be red and sore and I'd have to pull them away. For Leemo this was just a game and she'd leap into my lap clasping her big soft paws around my neck and licking my face. She was an incredibly loving animal and craved company. Yet there were only two friends in her life, Beano the ocelot and myself. No other human could go near her, and every other beast was either, like Chico, something to avoid, or like the livestock, something to pursue. Fortunately, the ranch animals usually gave the yard a very wide berth, the strong smell of jaguar and puma undoubtedly being the reason. Still, as could be expected, there were incidents.

Leemo and Beano had a favourite game that they played every morning in the shrubbery. It was just in front of Jimmy's house which threw a long shadow at that time of day and kept it beautifully cool. They had their own version of hide-and-seek; only Beano, rather reluctantly, did most of the hiding, and Leemo all of the seeking. It was a rough and tumble between an unmatched pair, and small wonder that Beano ever wanted to play at all. He would trot off and lie down behind one of the bushes; but Leemo cheated atrociously and always watched to see exactly where he was going. No sooner had Beano hidden himself, than Leemo would come bounding through the flower beds and jump on top of him with all her weight, which was nearly five times as much as his. The game rarely lasted more than five minutes, after which Beano would pick himself up and dash as hard as he could down the pathway for the safety of the den. Unfortunately for him the exuberant Leemo was not to be shaken off so easily, and invariably bowled him over at least twice *en route* before he could get to the house.

I used to watch these antics with quiet amusement from a chair on the veranda, as there was really no satisfactory way I knew of to chastise a full-grown mountain lion for being a bully. However, sometimes nature can even out the score, in interesting and subtle ways.

One day Leemo had just caught Beano a really hefty blow which sent him rolling over and over across the lawn. I was on the point of intervening when Leemo suddenly froze and looked intently across the garden and out through the fence. To my horror a small party of sheep were approaching from the shelter of the mango trees, not fifty yards distant. Before I could take any evasive action Leemo had sprung onto one of the fence posts and dropped softly down to the other side. The grass was very short, perhaps barely 6 inches in height. Yet this barely existent cover all at once seemed to conceal her adequately as she dragged herself flat on her belly towards the flock. There was one very tiny lamb amongst the group. It was as white as driven snow, save for the coal-black tip of its nose. Only a few days old, what did it know of the dangers to be found so close to home? It bounced up and down on its four little feet, wringing its tail; stopping every few moments to nibble at a shoot of grass, or to butt its mother impatiently in the breast before taking a few hasty tugs at her nipple.

The older ewes had already spotted Leemo and stopped in their tracks. It was as if they were rooted to the spot under a hypnotic stare. Instead of running while they yet had a chance, they watched and waited while she inched her way methodically towards them. Suddenly Leemo broke ground in a tremendous rush. From zero to 50 miles an hour in the space of a few yards. The ewes bolted like stung rabbits with the little white lamb only a pace behind. Leemo could have caught the leader in a few easy strides; but she had her eye on the lamb from the start; and the ABC of hunting, that all cats learn by instinct, is never to change your target once the chase is on.

The puma's kill is always a studied masterpiece in its execution. Not the power of the lion or the crushing strength of the jaguar, but a smooth, graceful, co-ordinated, deadly rhythm. Leemo drew level with the lamb's fleeting hindquarters. For a moment she was running on three legs, with one fore-paw poised. She waited for that precise moment when the lamb's near hind leg had reached its furthest backward motion. Then she tapped it, almost casually with her paw. The back legs crossed, the lamb tripped for an instant, but it was enough. Leemo caught its throat in her jaws, the pair rolled and the end was only seconds away.

The last thing Leemo, or I, or even another sheep would have expected was retaliation from the rest of the flock. Yet, as the lamb gave its first pitiful bleat of alarm the ewes skidded to a halt, turned, and charged *en masse*. Despite the incredible strength and ferocity of the cat family, their business is attack; when the tables are turned, as they sometimes are by an aggressive dog, they will usually flee if they can, rather than engage in a face-to-face battle, even one they could easily win.

So unprepared was she for such a development, and so engrossed in her prey, that she was taken completely by surprise. Four big ewes bumped their heads simultaneously against her rump with devastating effect. She was bowled completely over and almost before she regained her feet they hit her again. Completely put to rout, Leemo bolted for the garden fence. Once in her stride she quickly outdistanced the sheep, but it was a rather frantic and ungraceful leap that took her safely to the top of one of the posts. The ewes in a conclusive show of bravery, to the everlasting credit of their much-maligned species, pressed their attack to the very foot of the garden fence, while the lamb, none the worse for its ordeal, ran hurriedly off to the safety of the distant corral. There they stood, heads shaking defiantly, and a much subdued and shameful Leemo dropped meekly down to her side of the garden, and trotted painfully away in search of Beano and games she could be sure she'd win.

The residents at Dadanawa, beasts of fur and feather, as well as human, were however not the only ones to be harassed by Leemo. For instance, up to this point there had always been at least three reasons why the people who owned the ranch rarely visited it: first of all, because it was so difficult to get to, secondly, because when they got there the conditions were in their opinion primitive, and thirdly, the ranch paid such excellent dividends that it really didn't justify an awful lot of attention. There was very shortly to be a fourth reason.

About 200 yards from my house was a very attractive wooden bungalow which directors and other V.I.P.s used whenever they were in residence. On this particular occasion a senior member of the board was spending a couple of days on a tour of inspection, which was largely comprised of quail shoots in the morning, siestas

(*right*) The Wai-Wai hunting dogs are tethered on a raised platform. (*below*) Painting arrow heads with curare

A Wai-Wai draws his 8-foot bow

A successful Wai-Wai hunter returns with a tapir

(*above*) The secret of sleeping in a hammock is to lie across it

(*below*) Wai-Wai women spinning cotton

in the afternoon, and duck hunting and cattle viewing in the evening.

Leemo, unfortunately, did not see this as a time to be on her best behaviour. Instead she chose one after lunch period as the moment to jump over the fence and pursue three rather ladylike and aged milk cows mercilessly across the savannah. This would not have been a disaster had she chased them in any direction other than the guest house, which was surrounded by a very neat little paddock. When Jimmy had designed and built it, out of light wire, the object had been more to enhance the beauty of the carefully nurtured shrubs that it guarded, than to repel a stampeding herd of cattle. The sound was akin to a million snapping guitar strings when the terrified animals hit it square amidships. All three ended up with their heels in the air in the middle of the bougainvilia, but lost nothing in regaining their feet to sprint across the lawn and demolish the opposite fence with equal efficiency. By the time this certain grey-haired gentleman appeared at his front door, there was nothing left to see save a flattened fence, and uprooted flower-beds, the ploughed-up lawn, and a solitary mountain lion. For those who may think the misfortunes were not yet complete, there emerged at this juncture from inside the bungalow, the executive's own imported hunting dog—a wire-haired terrier named Anna.

Leemo, meanwhile, had lost interest in further pursuit of the cattle and began to lope quickly back in the direction of home. This movement was, unfortunately, construed by Anna as an act of retreat, and with a yelp she took up the chase, urged on by the proud encouragement of her master. The high-pitched bark of the rapidly approaching terrier soon caused Leemo to stop and turn around. Had the dog continued the attack, she might well have put her up the nearest tree, but instead, when she saw herself face to face with the business end of a very large cat, she came to a sudden halt. Leemo advanced with an angry snarl, and Anna lost her nerve. Whether or not she heard her owner's frantic command to heel, had little bearing on her amazing display of greyhound qualities. Even had Leemo tried, it seemed doubtful that she could have caught her during so hasty a dash for the safety of the bungalow, and instead she walked arrogantly away to take a quiet nap in the shade of some nearby cashew trees.

When a day or two later, Beano climbed in through the bathroom window and absconded with the gentleman's abdominal support, and his wife barely survived a charge from Chico by the breadth of her panty hose, it seemed generally agreed that director's visits should be restricted to those prompted by reasons of great importance.*

* For the complete adventures of Leemo and co, see *Leemo*, and *More about Leemo*, by the same author.

XII

ZERO-TWO-ZULU AND THE WAI-WAI INDIANS

During the next few months two important changes took place. I was promoted to the rank of general manager and, in January 1965, was given an aeroplane. The first was of minor significance. Jimmy Brown didn't want the job and continued to be in sole charge of construction and demolition. It was merely a new title for my old post, but, although my duties remained the same, I felt I had gained a certain amount of prestige. On the other hand, the advent of the aircraft was probably the paramount change in administration since the ranch swung from private hands to company ownership in 1920.

Overnight, journeys of four days became jaunts of only ten minutes. Far-flung outposts and corrals scattered strategically about the ranch for round-up purposes, and completely inaccessable throughout the wet season, could be reached during the highest floods. Airstrips were quickly made, until a total of twelve were spotted across the range. Outbreaks of disease were readily detected from the air. Wherever there was a dead animal there would be a flock of black vultures, circling or perched in trees. Nothing from the bird's-eye view of an aeroplane could be more obvious. In one day's flying the entire 3,000-square-mile territory was easily patrolled, while sick or injured Indians, who traditionally fended for themselves or sought cure via local witchdoctors, were flown to Georgetown for medical aid.

The tiny, single-engine aeroplane that made all this possible was a red and white Tri-Pacer. The registration lettering Zero-Two-Zulu became its affectionate name. It had been modified for use on short, rough jungle fields by the original owners, a missionary group. This very simply had entailed the removal of the nose wheel and the positioning of the main landing gear further forward with the addition of a tail wheel. This alteration had achieved two things. First, it removed the stress on the comparatively less robust nose wheel which resulted during bumpy landings and take-offs; and secondly, it removed the quiet, responsive ground-handling capabilities that are characteristic of planes with steerable nose-wheels. The first result was for the better, the second was an abomination. Zero-Two-Zulu, in polite aviation vernacular was a ground-looping son-of-a-bitch.

After two weeks of instruction at the Georgetown airport, and forty hours experience, I flew across the 350 miles of forest to Dadanawa. The flight took three and a half hours, and never before had the dense canopy of leaves looked so ominous and void of clearings.

The day after my arrival at the ranch I flew to each of the newly constructed airfields. They averaged 900 feet in length and were mostly built on the tops of hills, well clear of areas that might be flooded or soggy during wet weather. The work involved in making a Rupununi airstrip was minimal. Essential requirements were a combination of firm ground, reasonably level terrain, clear of trees and boulders, and heading into the wind. Two of these were considered adequate. So we had strips on firm ground that were clear of trees, but littered with boulders and in a cross-wind configuration; and we had strips on firm ground clear of boulders that had such a hill in the middle that you couldn't see the far end. After selecting a spot, four Indians were then assigned to the task of hoeing the worst bumps off and digging out tree stumps and boulders. The clearing of a strip 900 feet long by 20 wide usually was accomplished in a week.

I had inspected five of the airfields while the Indians at each stood by in complete awe. For many of them it was the first time they had seen an aeroplane, other than those that occasionally passed overhead at 37,000 feet. Being able to actually drive one of these amazing contraptions, no doubt shoved me up in their estimation. The next field on my list was at a small outpost named Weri-More, tucked away at the foot of the mountains. I approached and flew along it a couple of times to make sure that everything appeared to be in order. It lay parallel to a wooded river bank on flat ground and headed nicely into the wind. The only obstruction was a lake at the far end of the strip. Four Indians stood proudly by with hoes on their shoulders, like rifles at slope arms. I let the flaps down, reduced power and began my approach.

Looking back, I like to think that the length of the field was entirely responsible for the mishap. Indeed, it turned out to be considerably less than the original measured 1,000 feet. But with only forty hours experience, a pilot is apt to make the wrong decisions. That day I should have hit the throttle and 'gone round again'. But I didn't, I landed; and it was too late when I finally realized I was going to run out of field, and turn into a duck.

As the water rushed towards me I stood on the brakes for all I was worth. Alas, without the nosewheel, Zero-Two-Zulu couldn't stand that kind of treatment. She pitched forward on her nose and somersaulted onto her back. Suddenly, I was hanging upside-down from the shoulder harness in dripping fuel. Dazed, I felt for the master switch and turned it off. Then, without thinking, pulled the release on the safety belt and fell on my head. I lay there for a moment resting on the top of my shoulders, neck crumpled, peering between my knees. Slowly, I wriggled over and crawled out through the shattered windshield.

I emerged from the debris, stood up, and found myself facing the four Indians, still carrying their hoes at the slope. They smiled and the foreman stretched out his hand to shake mine in customary greeting.

"*Ai, kaimen pouga?*" he said cheerfully, which translated into English means very simply, 'Good morning, how are you?'

They thought that this was the normal way of landing!

It cost 2,000 dollars to repair and a note of anguish from the insurance company. But within four weeks I was airborne once more and, needless to say, never made the same mistake again.

A hundred and ten miles south of Dadanawa lived a very interesting tribe of Indians known as the Wai-Wai. I had visited them on a number of occasions, but, as the entire distance was through dense untracked forest, it was an extremely arduous and time-consuming operation. There were about 300 Wai-Wais, and they lived at the village of Kanashen in the headwaters of the Essequibo. Originally, they were a Brazilian tribe from the Rio Alto Trombetas, a northern tributary of the Amazon. Inter-tribal wars, common amongst Amazon tribes, had driven them across the watershed into Guyana.

The Wai-Wai have no nearby neighbours to the north, since they are isolated from the Wapishanas by the forest, and there has been little contact between the two except through trading. Their main connections, therefore, are with the primitive tribes to the south, across the Serra Acarai mountains which form the divide between the Essequibo and the Amazon. Here live the Mawayanas, Kat-y-ana, Piskaryenna, and Parukotos. There has been considerable inter-marriage between the Wai-Wai and these Brazilian tribes, which is fortunate, as without it they would probably have become extinct. It is generally believed that the Wai-Wai were almost entirely wiped out in a massacre led by the fierce Paricote Indians sometime in the 1890s and a mere handful of survivors fled north over the Serra Acarai.

The Wai-Wai are the most primitive of all the Indian tribes found in Guyana, a fact due entirely to the impenetrable terrain in which they live. Actually, the only noticeable change in their culture is the disappearance of their stone hatchets, knives and other implements. These have now been replaced by the metal variety which they obtain from the Wapishanas in exchange for hunting dogs, bows and arrows and other products of Wai-Wai expertise.

I had become interested in them as a result of a programme of mammal collecting that I was conducting for the Royal Ontario Museum in Toronto. We were mainly interested in bats, and as the upper Essequibo was largely unexplored, I had made a number of collecting expeditions there in hopes of finding new species. Now, with the aircraft, this work was greatly simplified, for there was a small airstrip at Kanashen that had been built by a group of American missionaries.

Hovering only 5 miles an hour above stall speed, I nursed Zero-Two-Zulu along at tree-top level. Foliage from Giant Protiums and Turu Palms raced past the window as I lowered the last notch of flaps. Suddenly the trees ended, and there was the strip, a hand-hewn clearing in the forest rushing towards me with awesome rapidity. Wheels touched hard clay with a screech, then a bump, and I was jerked forward against the shoulder harness. We bounced twice more and I pulled the stick hard back into my belly. Half the field was gone and a wall of jungle at the far end loomed dangerously close. I hit the brakes as hard as I dared and the aircraft vibrated like a plucked guitar string. My chin hit me in the chest, and then, with a final lurch, we came to a halt, twenty paces short of the first trees. I looked out of the window and breathed a sigh of relief. Moments later twenty or thirty scarlet-painted Indians carrying enormous bows stepped into the clearing.

As I got out of the cockpit several of the Wai-Wais came forward to meet me. They were fine-looking people, broad-shouldered, deep-chested, and about an average of 5 feet 7 inches in height. My first impression was the unusually light tone to their skins. Then at once I realized why the Wapishanas called them Wai-Wai, for *Wai-Wai* is the Wapishana for tapioca, a product of bitter manioc, and of course very white in texture. Here in the tall rain forest, with 120 inches of precipitation a year, more than double that of the savannahs, people are protected from the bronzing effects of the sun.

The first Indian to approach was one of the headmen, Fon-yuwe, who had helped me on my previous trips. I shook hands with him and then with each of the others in turn. The men wore a loin cloth of red cotton, while the women, who had just begun to venture out to the edge of the clearing, wore a small rectangular apron woven entirely from strung beads. Each man had his face painted in an individual design using red and yellow dye derived from anatto seeds (*Bixa orellana*), and the hair was worn waist length. It was their hair, more than anything, which made them so distinctive from all the other tribes of the northern Amazon basin. Instead of being loose, falling casually upon the shoulders like the women of the tribe, it was gathered into a long pigtail, bound tightly with fibre, and then thrust inside a hollow bamboo tube some 18 inches in length. The tube, which dangled at their backs, was decorated at the base with clusters of toucan feathers. The entire hair-do took on a rich glossy sheen, being well greased in monkey fat. It was cut across the forehead in a fringe, and some of the men had scattered small fragments of pure white down, from the breast plumage of a harpy eagle, about their crown. At a glance, the hair appeared to be flecked with cotton wool.

A Wai-Wai's worth is measured in the number of beads he owns. They are gathered on long strings, often measuring as much as 50 feet. Most of the men wore some around the neck, but the majority were wrapped in tight bands just below the knees, or on the arms between the biceps and shoulder.

Everyone in the group carried a bow slightly more than 8 feet in length, and a bundle of twelve to fifteen arrows. None of the arrows measured less than 5 feet and they were made from the straight slender stems of arrow-grass cane (*Gynerium saggittatum*). The reason for the incredible length of the bow is merely to provide sufficient power to send the arrow 200 feet into the tallest tree at a monkey or edible bird. The pull is between 70 and 80 pounds. Arrows are made long so that when a shot is missed and falls far into the jungle, it can be found because the shaft sticks up above the tall undergrowth. All arrows were bi-fletched, with feathers a full inch broad, made from the tail of a curassow, a large turkey-like forest bird. They had a natural curve and were set so that one curved to the right and the other to the left. Hand-spun cotton bound each to the shaft in an intricate manner. Both ends of the binding were decorated with short orange tufts of toucan feathers. The width of the fletches helped to stabilize the long, heavy pointed arrow. Because cane is hollow, the nock is carved from hardwood and inserted like a plug into the tail piece and tied in position. They made two distinct types of points. One was hardwood, long and slender, with a serrated edge like a saw, and used on monkeys and birds. The other was bamboo, flat and dagger-shaped, for big game such as deer and tapir.

Fon-yuwe was a particularly well-built man, probably in his early forties. It is extremely difficult to estimate the age of any Indian between 25 and 45, as they seem to change very little in appearance during those years. He had a perpetual smile on his face, a feature which endeared him to me greatly. Never once did I have the feeling that he resented my presence, as I had during my early contact with the Wapishanas. Fon-yuwe had proved on my previous expeditions that his knowledge of the forest fauna seemed quite without bounds. In addition, he spoke reasonable Wapishana, which he had learned from a solitary man of the latter tribe who, for some obscure reason, had migrated to the Essequibo and taken a Wai-Wai wife. As a result, we communicated in admirable fashion and Fon-yuwe was indispensable. The value of such a friendly relationship cannot be over emphasized. All forest-dwelling Indians have a definite tendency towards isolation. Their lives depend on hunting, and the hunting deteriorates when too many people begin to congregate. This had bred in them an independent mentality which makes them feel uneasy, perhaps that their liberty is threatened, whenever outsiders arrive in the village.

I followed Fon-yuwe along a narrow pathway which led to the village some 200 yards distant. The houses were round with conical roofing of palm-leaf thatch. The thatch came to within 4 feet of the ground, and the walls were merely a barricade of close-standing sticks, with one gap left for the entrance. Fon-yuwe took me to his own hut and we went inside.

The atmosphere was thick and smoky and as soon as I stepped through the door several dogs immediately broke into an angry violent fit of barking. I stood still for a moment until my eyes became accustomed to the dim light lest I inadvertently walk into one and be bitten. The animals were standing on wooden tables built around the perimeter of the walls, 3 feet or so above the baked mud floor. These were hunting dogs and were favoured with very special, yet sometimes brutal, attention from the women of the house. Wooden rods 15 inches in length were attached directly to their collars while the other end was connected to a rawhide lead. The wood prevented the animal from getting free by chewing its way through the hide. The Wai-Wai are famous for their hunting dogs, which are valuable trade items in inter-tribal commerce. They are extremely bad tempered and used mainly to bay large game such as deer, tapir and peccary. There were seven of them in Fon-yuwe's house and they flung themselves furiously at the ends of their leads as I approached. Immediately, one of the women picked up a large stick and beat them mercilessly over the head until each one cowered on its table snarling fiercely.

Fon-yuwe's family seemed to consist of two elderly women and two young girls. Except to belabour the dogs, none appeared to take the slightest notice of our entrance but continued with their work.

They were making farine, which is the Wai-Wai's staple food as it is for the Wapishana. Farine is manufactured from the root of the bitter cassava plant, sometimes called manioc, which in its unprocessed form is deadly poisonous. The tubers reach maturity in twelve months, when one plant will produce as many as six radiating roots each as thick as a man's arm and 18 inches in length. As soon as the full-grown plants have been pulled out and the tubers removed, the stalks are then cut up and replanted three or four to a hole, for next year's crop. This is done just prior to the wet season, but any care thereafter is left entirely to nature.

After the harvest, the roots are soaked in water for three days to soften. Next, they are mixed with a similar quantity of dry, unsoaked tubers, and grated. As we entered Fon-yuwe's household, one of the young girls was doing this. She utilized a wooden board about 15 inches long and 10 inches wide, into the face of which had been planted several hundred tiny fragments of rock. Each individual chip had been set by hand, tapping it in place with a stone hammer. The Wai-Wai are so expert in the manufacture of these graters that they have long been the tribe's chief export item to the Wapishanas.

After grating, the powdered cassava is forcibly stuffed into a 9-foot-long cylindrical resilient tube plaited from palm leaf, known as a matapee. During this process the tube becomes greatly shortened and widened. At the top of the matapee there is a loop and this is threaded through a rafter in the house roof so that the apparatus hangs well clear of the ground. A pole is then passed through another loop, connected at the bottom, for about two-thirds of its length. The shorter end is wedged into a slot in a peg firmly driven into the ground. This results in a see-saw type lever which the two older women were in the course of manipulating. As they forced downwards on the pole the matapee yielded and began to elongate and compress. All the watery and poisonous contents of the grated fibre now became expressed through the bottom of the matapee and were collected in a large pot. This juice is then boiled over and over again until the poisonous elements evaporate and a thick sauce known as cassarip is left. The technique is widespread amongst Indian tribes throughout the Guianas including the coastal belt. When the early Dutch settlers arrived in the seventeenth century they took this cassarip and used it as a base for a dish called 'pepper pot', which is still a favourite with natives throughout the Caribbean. After every meal, left over meat is tossed into the pepper pot cauldron, and periodically fresh cassarip is added. It acts as a preservative and the greater the age of a particular pot, the finer the taste. Families have been known to keep the same pepper pot replenished—and, of course, uncleaned—for as long as thirty years!

Another by-product of the process is our familiar tapioca. If the poisonous juices, before boiling, are allowed to settle, a white starch sediment forms on the bottom. The juices are poured off leaving only the starch, which is dried in the sun and then roasted in a pan.

After the two women had squeezed out all the poison from the matapee the mealy residue was dumped out and rubbed by hand through a basket-work sifter. Later it would be strewn out onto a large iron pan and baked over a fire. If farine was required, the powder would be stirred constantly until it hardened in an hour or two to a dry grit. This form is favoured by the Wapishanas. The Wai-Wai prefer to make it into cassava bread and the only difference in manufacture is that the powder is not stirred but allowed to bake into a large circular cake some half an inch thick.

How such a complicated procedure for eliminating the deadly juices from a poisonous plant was ever discovered, and could evolve into the staple food for generations of primitive man is inexplicable!

I spent the night in Fon-yuwe's hut, slinging my hammock across one corner, as far away from the growling dogs as I could get. The family slung theirs one above the other within a few inches of the fire which was kept burning until daybreak, the smoke filtering upwards through a central hole in the roof. In the morning we would move several miles down river to spend a day or two searching for new bat specimens.

The sun had barely cast its glint on the dark waters of the Essequibo when our 30-foot dugout canoe slid silently downstream from the Wai-Wai village. Sweat soon began to gleam on the broad naked backs of our four Indian paddlers. Only one of them was a full-blooded Wai-Wai, the rest were from other closely allied Brazilian tribes. Fon-yuwe was a Mawayana from the Rio Mapuera; his son, Too-e-foo, paddling alongside him, was by a Wai-Wai woman. He appeared to be about 17 years old. Pi-kocoo was Kat-y-ana, a tribe which lived three months paddling distance away on an upper tributary of the Rio Alto Trombetas. The fourth was a tall, lithe Wai-Wai youth named Te-tai.

As we rounded a bend and drifted by a shallow sand bar there was a sharp word of command from Pi-kocoo in the bow. At once we held water and the Kat-y-ana stepped quickly ashore, fitting an arrow into his bowstring. He slunk swiftly across the sand like a cat, with his weapon at half draw. In midstream a gentle V-shaped ripple cut across the surface. Suddenly Pi-kocoo came to his full height and 100 inches of seasoned wood arched effortlessly. There was no ponder of aim, the string was released the instant full draw was attained. A 7-pound lukanani, arrowed behind the gills, seemed a foregone conclusion to such a beautifully co-ordinated movement.

A mile further on the Indians began to slap their paddles rhythmically against the sides of the dugout with every stroke. They lunged at the water with such united force that the bow wave rippled level with the gunwale. At the next bend a column of smoke arose through the trees. We were approaching Onoro village.

The noise of the paddlers soon brought a party of Wai-Wai to the river bank. We pulled into shore amongst several other canoes, and made fast with a rope of green twisted wood bark.

I questioned Fon-yuwe about the stop, for my time was limited, and had no wish to spend it socializing with his neighbours.

"Onoro, wom turryan urari!" he replied quickly. This was the home of the local curare factory, and they wanted to buy a quantity of poisoned arrowheads for use on the trip.

We had to walk several hundred paces to the village, followed by a dozen children who tugged curiously at my clothes. Fon-yuwe led us to an open-sided palm-roofed shelter where a number of Indians were squatted around a fire. There was no customary greeting from any of them. All were absorbed in the activity of one central figure, a middle-aged man surrounded by scores of arrow points, arranged neatly on frames, stirring a red bubbling liquid in an earthen pot.

I was very excited at being able to witness the actual manufacture of the poison, and asked Fon-yuwe to explain the process. It runs as follows.

The important base is the root of a large forest liana, called by the Wai-Wai, *ba-la-wi-tyu*, which is in fact the strychnos vine. It usually grows on mountain slopes, where the root is dug up and cut into short lengths. Later this is chopped into shavings and soaked in water. These shavings are then squeezed and strained through the hands producing a red-brown liquid which is accumulated in an earthenware pot.

Various other ingredients are added, although it is doubtful that they add much, if any, potency to the preparation. Among these are the bodies of tarantula spiders, stinging ants, and red peppers. The next addition are the leaves of a small bush called kunan (*Clibadium sylvestre*), which is widely used by many Indian tribes as mild fish poison (if the leaves are pounded up and thrown into the water, they kill a variety of small fish up to about a pound in weight). However, despite the toxic elements of kunan, the strychnos is really the fatal component.

The mixture is then boiled from sunset to sunrise, during which time it must be stirred constantly. Though at this point near completion, the concoction still lacks the necessary viscidity to cause it to adhere to the arrowhead. This is provided by the stalks of a dwarf palm which secretes a glutinous sap. A bunch of these stalks are dipped in the mixture, then wrung through the hands extracting the sap and blending it with the poison to produce the required gumminess.

The final stage is the application of the curare to the arrow points. Points thus treated are not permanently attached to the shafts, but carried separately in small quivers. They are slivers of bamboo, 9 inches long with the flat surface along one face faintly notched with a sharp knife at one-eighth of an inch intervals. The poison tends to lie in these grooves. The curare is applied with a fibre brush to about twenty points at once, which are held by inserting their butts into holes bored in an old arrow shaft. Part of the ritual requires that this arrow must be one that has killed an animal. The painted tips are then dried over the embers of a low fire, and the process is complete.

Making curare is a hallowed ceremony. The manufacturer must collect all the materials with his own hands, even the firewood. He must build his own fire, fetch his own water from the river, even make his own pot. Furthermore, no woman is allowed to witness its preparation, and he cannot make it if his own wife is pregnant. Any violation of these rules is believed to render the poison ineffective.

Curare is a remarkably stable product, and treated arrows, if kept dry, will maintain their deadly strength for years. Its effect is to paralyze the nerve centres, causing the muscles to relax and asphyxiation to take place as the lungs cease to operate. There is no effect on the meat which can be eaten safely. Monkeys are the main target for the curare arrow and, regardless of where the animal is wounded, death comes in several minutes. The only effective first aid for dealing with humans under the influence of the poison is artificial respiration.

Fon-yuwe and his three companions purchased a quantity of the treated arrow heads, which each man fitted into a small bamboo quiver wrapped in cloth. Payment was made in strips of dried peccary meat and cassava bread, while Pi-kocoo gave four newly-made arrows. The transaction completed, we left Onoro and were soon in the dugout and on our way again.

We beached the canoe on a lonely sandbank several miles below the village with the last rays of light. Working swiftly and expertly the Indians soon erected a palm-leaf shelter, while Fon-yuwe helped me to hang up the mist nets. Later, after a meal of roast fish and cassava bread, we slung our hammocks; sleeping by the friendly glow of a log fire, serenaded by the monotonous wail of howler monkeys, and the distant ominous grunt of a roaming jaguar.

At dawn there were several bats in the nets. We collected them and then started into the jungle on foot, for there were numerous caves and hollow trees in the area where additional specimens could easily be caught by day.

I followed the Indians along a barely discernible trail marked only periodically by a snapped twig at finger height. Fresh game spoor was evident within a few hundred yards of camp. Red and grey brocket deer, peccary, tapir, and at one spot a small newly-scraped heap of dirt and leaves where a jaguar had covered its droppings.

"*Kamara!*" Fon-yuwe hissed, pointing at the sign with the tip of his bow. But there was little chance of our meeting him, and we pushed on.

We waded a small creek and struggled through soft muck on the other bank. As we climbed gradually to higher ground, the air was filled with the shrill piping notes of white-throated toucan. Te-tai quickly strung his bow and fitted a bird arrow to the string. On the savannahs I never allowed my Indians to shoot toucan, but in the forest food is always scarce and the Wai-Wai depends on whatever he can lay hands on for survival. A toucan is not a large bird and the one Te-tai had spotted was on a wispy branch nodding in the breeze a good 120 feet above the ground. He stood with feet well apart and body gracefully arched to meet the almost perpendicular angle of the shot. The bow was grasped in only three fingers of his left hand, while the forefinger and thumb made a V over which he laid the arrow. Four finger tips of the right hand held the string below the nock, and the thumb kept a steadying pressure on the upper surface of the nock against the edge of the forefinger. He drew the giant bow until his hand came back to the point of his right shoulder. There was a noticeable swish as the 5-foot shaft grazed past the bow an instant after it was released. It soared almost casually aloft followed intently by hungry eyes. Only a slight cry came from the bird as the point took it an inch below the brisket; doomed, it toppled earthwards.

We hunted for several more hours in a semi-circular path towards camp. A number of caves were explored, and, by setting nets across the entrance and then frightening the occupants we were able to catch numerous bats. Other birds were also taken for the pot, and yet, although shots were missed, none of the arrows were lost. At first, it was difficult to understand how this was possible in such thick forest. Later I noticed that as an arrow went wide of its target the owner listened carefully for the sound of it striking foliage as it plunged into the undergrowth, and mentally marked the spot. On many occasions I saw them fire three and four arrows, all of which missed, and then run into the forest and retrieve every one in the space of five minutes!

That night I was plagued by an acute irritation in my feet, and a burning sensation around my midrift. Either one would have kept me awake, but the two together were sheer misery. I knew exactly what both of them were, for they occurred occasionally on the savannahs. My feet were infested with chigoes and my middle was host for bête rouge.

Chigoes, or sand fleas, live in dry sandy interiors of native dwellings. Damp discourages them so they are rarely found in the open. Obviously Fon-yuwe's hut had been full of the pests. They burrow under the skin, usually on the soles of the feet or under the toe nails. It takes about two days for the insect to make its presence known, which it does with a severe itch and bluish spot on the skin. This is the capsule which generally contains hundreds of eggs. If allowed to hatch the worm-like maggots slip out and continue the boring operation, quickly spreading over a wide area. The only remedy is prompt removal, an operation that all Indians are expert at.

Too-e-foo performed the service using a sharpened splinter of hardwood. He picked up and turned aside the skin surrounding each capsule and carefully lifted it out in its entirety. This left a neat crater-shaped hole the size of a small pea which quickly heals. Altogether Too-e-foo removed fifteen from the soles of my feet.

The bête rouge was a separate problem, and unfortunately quite without a magic cure. These almost microscopic insects lie in wait on blades of grass and attach themselves to passers-by. Like the chigoes, I had collected them at Kanashen, where long grass had sprung up around the village clearing. The insects make their way up to the victim's groin and belt region, and burrow under the skin using their long proboscis. So attached, and invariably by the hundred, they give rise to an unbearable itch which leads to acute inflammation and small painful lumps. The burning sensation is so acute that the inclination to scratch is irresistible. At night it becomes intolerable and only when the body is completely exhausted is sleep possible. I have found that Acriflavin solution painted over the entire affected area will usually cure it. Unfortunately I had none with me. Left to run its course the insects seem to drop off after three or four days. But in areas where bête rouge occurs constant reinfestation can make life virtually insufferable.

So it was after a restless night that we paddled downstream next morning to hunt a new area. After a mile or two we beached the canoe and split up into two parties. Fon-yuwe and I went off to search for more bats, while the other Indians were left to hunt for meat to take back to Kanashen.

Fon-yuwe was without a doubt the most competent man I had ever followed into the forest. He could leave camp in the morning, walk for 20 miles in a circle and arrive back at camp again just before sundown. Whereas most Indians safeguard themselves against getting lost by blazing one trail in the morning and returning via the same route in the afternoon. The magnitude of this can only be appreciated by someone who has seen an equatorial jungle and realizes how dense and trackless it can be. Thousands of trees begin to look similar, there is no horizon and the direction of the sun is only occasionally apparent through the tall forest. One can become totally confused and then hopelessly lost within a few yards of a familiar trail. The bones of many men so doomed lie buried in the mossy undergrowth of the South American selvas.

The advantages of a guide like Fon-yuwe on an expedition are great, because the same ground need not be covered twice. By mid-afternoon we were nearing camp and still finding new bat roosts. Our collecting bags were full.

Suddenly, as we were crossing a small gully there was a shrill whistle close by. At first I thought that we must have come up with

(*right*) A Wai-Wai woman making a cassava grater.
(*below*) Making farine from bitter cassava

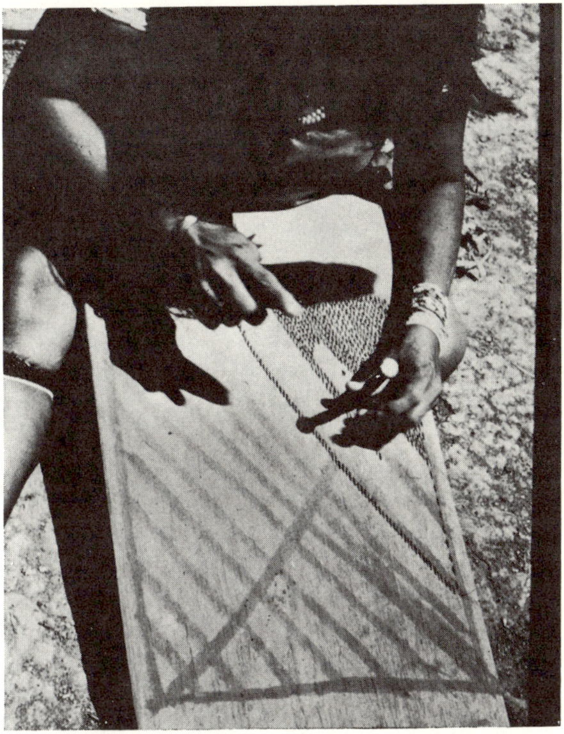

(*left*) Trio Indians

(*above*) Trio women along the trail to the Paroe River

(*overleaf*) A Trio Indian along the trail to the Paroe River

the other Indians, and would have replied had Fon-yuwe not clutched my arm excitedly.

"*Ya-y-feu* [tapir]!" he hissed. "*Mashovenry um dakadanna* [wait, I will call him]."

This was not the first time I had heard a tapir whistle, but being close to camp and so human-like in tone, I was quite deceived by it.

Fon-yuwe put two fingers in his mouth and gave one short blast. It was answered immediately. He whistled again, and this time the reply was accompanied by breaking branches and bursting lianas as the largest and most powerful beast in South America came lumbering and barging towards us.

Fon-yuwe quickly removed a curare point from his quiver and fitted it to an arrow. He was fully drawn and waiting when the tapir pounded into view less than a dozen paces away. At such short range the arrow struck with the full force of the enormous bow and was buried deep into the point of the shoulder. The animal careered sideways and smashed the shaft against a tree, then staggered forward and reeled drunkenly for 50 yards. The Mawayana gave a triumphant shout, danced forward and drove a second bamboo blade through its heart from the distance of only a few feet. Death would hardly have been quicker with a rifle bullet.

After removing the guts and head, the meat and hide still weighed in the region of 300 pounds. Fon-yuwe plaited a backpack from a large palm leaf and we lashed the carcase securely into it. The only assistance he would accept was help in lifting the load into position on his shoulders. It was an unenviable 2 miles to camp, which he completed with only two rests. In a society that must kill in order to survive, Fon-yuwe would be a hero for many days.

The other Indians had also been fortunate and killed a grey brocket deer. That night all the meat was smoked over a large barbecue to preserve it during the return journey to Kanashen.

Next morning we started the long paddle upstream. It would take us several hours, for the current in the Essequibo was strong and our dugout heavy from the meat. The task was made easier by the fact that we were returning home after a successful hunt. I had my bat specimens and the Wai-Wais had their meat which they chewed away at constantly as they strove with the paddles.

Throughout the day a low rumbling of thunder to the east and a steady build-up of towering cumulus clouds warned us of an approaching storm. By mid-afternoon it broke and our little canoe was soon being thrashed under a violent deluge. It rained so heavily that at times we were unable to see the river bank only a dozen yards away. Lightning cut the air all around us, and it was soon essential for one man to abandon paddling to bale out the gallons of rainwater that threatened to sink us. Only by working furiously with the paddles could we keep our bodies from freezing, and my legs and groin became seized with cramp.

Finally darkness came and, after a battering that had lasted for almost four hours, the rain gradually ceased. At this point we were still a few miles from Kanashen, but close to Onoro village where we had collected our supply of curare. We could see a number of fires glowing in the distance, and after a short discussion decided unanimously to pull into shore and warm ourselves up.

As we arrived at the doorway to the first house our greetings were drowned by a savage chorus of barking dogs leaping at us from their elevated perches around the walls. My Wai-Wai companions pushed their way through the entrance without hesitation. I followed immediately, my desire for the warm interior quickly overcoming my fear of the mongrels.

There was a huge log fire burning in the centre of the room. I took my place with the others and squatted, shivering, on the mud floor only inches from it. A large pot hung over the flames on a pole set between two forked sticks. The contents bubbled furiously; foamy light-brown liquid oozed over the edges like dirty soap suds, and fell hissing into the embers. An elderly woman bent over the earthenware cauldron and churned over its contents with a long stick. She wore only a scant bead apron and her long hair and elongated breasts hovered only inches above the gurgling stew.

"*At-ana,*" she said, with a satisfied nod of the head.

Pi-kocoo was the first to reach forward. "*Fo-ro-to* [spider monkey]," he murmured. He dipped his hand quickly into the pot, seizing a steaming portion in his fingernails. "Ah!" he exclaimed. He dropped the piece to the floor and shook his burned fingers. Then he grinned and picked it up. It was the complete limb and hand of the monkey. He took a bite and handed it to me.

"*Ches-eres-mareu* [let's eat]!" he offered politely.

My hands were still partly frozen and I scarcely noticed the heat on my fingers. I wrenched off a piece of the rubbery flesh with my teeth, chewed it twice and swallowed it whole.

The family watched with satisfaction as we all ate ravenously. There seemed to be seven people living there. One was a very old man, his face deeply lined, and his chest intricately carved in a network of fine scars, marks of many ritual blood-lettings in the interest of dispelling evil spirits. The elderly woman was probably the wife, and there was another couple, perhaps the man's son and his wife. They appeared to be in their middle twenties, and the young man joined in the meal while talking rapidly with Fon-yuwe and the others. Three children, two boys and a girl, sat on a dry deerhide, several feet from the main group watching the proceedings quietly but intently. They were between 7 and 12 years of age, and with their shoulder-length hair, it would have been difficult for an untrained eye to have known which was which. None wore the customary beads about their arms which meant that they had not yet reached puberty. Only then would they be permitted to adorn themselves with the 'apomi', the band of white beads worn tightly below the shoulder; and a boy could wear his hair in the pigtail and bamboo tube. At this age sometimes only the girl's bead lap, and the piece of red cloth tied around a boy's scrotum, could be used by a stranger to tell them apart.

Meanwhile, my Wai-Wai companions were making short work of the pot full of meat. I had finished my piece of forearm and found it palatable enough, and the hot juices very welcome in my chilled stomach. We had mixed it with large pieces of baked cassava bread, wetting them first in the thin gravy and it amounted to more than a substantial meal. Suddenly Too-e-foo, who had been rummaging around in the bowels of the pot pulled out his hand with a cry of triumph. I made the mistake of looking at him with a smile, for he immediately offered me the unfortunate monkey's skull, still covered in black shaggy hair.

"Kai-men!" he exclaimed, pouting his lips, 'this is the best part'. I accepted the object graciously, not wishing to offend the household with a refusal. I ate one cheek and an ear, then handed it back with a nod of agreement. It was two or three minutes before I could work out all the strands of wiry hair from between my teeth; it's amazing what hunger, and the desire to remain polite will do to a man.

After half an hour our bellies were full and the circulation returned to normal. Fon-yuwe went back to the canoe and brought a large portion of smoked tapir meat which he presented to the old man. This done he turned to me, and with a sweep of his arm announced, "*Kayka-ha!*" My sparse knowledge of the Wai-Wai language interpreted this as the signal to go. "*Wo-mukunna,*" I echoed in Wapishana. Somewhat reluctantly we left the friendly hospitality of the Onoro village and serenaded by a thousand bull frogs, and the melancholy wail of a distant howler monkey, paddled the last few miles upstream to Kanashen.

XIII

MORE ANIMAL MISCHIEF

Next day, I said good-bye to Fon-yuwe and my Wai-Wai friends and flew back to Dadanawa. I was sorry to leave them, but after four days I was growing apprehensive about what mischief my animals might have been up to during my absence.

"They does know when you ain't here," said Irene, facing me with hands on hips. "God knows what would 'ave 'appened if you 'adn't locked up Leemo an' Chico. But it been bad enough with the others!"

I listened to a long list of unfortunate escapades that one could hardly believe occurred in such a short space of time.

The police had been to the ranch, on a routine patrol, and as usual stopped for a meal. Apparently one of them fancied himself as a bit of a cowboy, and seeing a lariat hanging on the wall removed it to give a demonstration. Unfortunately Tiny, now 500 pounds of incredibly strong tapir chose that moment to amble by. Well, a tapir is a myopic creature, relying entirely on an acute sense of smell and hearing for navigation. When alarmed his only defence is instant flight, blundering through the forest with the speed of a horse and ferocity of a rhinoceros. Nothing stands in his path. Tiny's mind was far away, probably in the banana patch, and the last thing he expected was an ambush. Suddenly from nowhere there was a hoot from Dadanawa's newest rodeo star, and a smart thwack around his shoulders. Tiny felt a jerk as the noose tightened about him. He was trapped! At once his brain switched off, except for one small transmission to his legs, which said, 'run like hell'. With a squeal Tiny dropped into high gear and arrowed his way across the yard. The policeman made one vain attempt to stop him, but dropped the rope with a howl as rawhide seared through his hands like a red-hot poker.

Had Tiny been in the open he would probably have run for a few hundred yards and then stopped. Regrettably he was not, and

Jacko's pen was scarcely an adequate barrier. The tapir hit the wire-mesh door head on, and carried it through the rear wall with hardly a check in speed. Both monkeys hit the roof with stark terror and instant diarrhoea. When the tornado had passed they stampeded through the hole and headed for the mango tree.

Unfortunately, within an hour or two they recovered from the shock and soon set about their usual rampages.

Big Jacko got upstairs in the house and dumped a pile of my best reference books into the toilet bowl. Luckily he didn't discover how to pull the plug, but before departing squeezed the entire contents of a tube of toothpaste over the floor.

Meanwhile, out in the kitchen, Little Jacko managed to steal two dozen plastic clothes pegs, and ate all the Neocide fly paper. The incredible thing about the latter was that the papers were clearly marked, "Poison, Do not Swallow," and in past years had caused the death of four domestic cats and one marmoset. Apparently spider monkeys have some inherent medical knowledge. After eating the Neocide he raided the medicine cabinet, and chewed up a whole box of Aureomycin soluble oblets, a powerful antibiotic used in the treatment of bowel disorders in bovines!

Fortunately their liberty came to an abrupt end when Jimmy caught one of them running across the yard dragging a pair of his khaki trousers. He promptly assigned four Indians to repair the monkey cage and by mid-afternoon the two delinquents were once again safely under lock and key.

Although reluctant to subject the Dadanawa staff to further harassment by my animals, it was time to release the cats. I turned Leemo out first, and, after showing appreciation by rubbing her sides against my legs, she ran off with Beano to take up sentry along the veranda wall. After they were safely gone, I let out Chico and Choo-Choo. Chico immediately went to his favourite spot beneath the orange trees, while Choo-Choo trotted off to the storeroom.

We had a margay cat living there which Choo-Choo was very fond of; a strange relationship akin to the one between Leemo and Beano. At the time Choo-Choo weighed about 40 pounds and found it impossible to follow the margay into some of the nooks and crannies of the storeroom that were his favourite lair. If she couldn't find him she meowed pitifully with a noise like a strangling alley cat, which had no effect on the margay who continued to hide contentedly on his own. When the sound became unbearable Irene would dig him out from behind the boxes and give him to Choo-Choo. This was quite a risky operation for as soon as the jaguar saw her friend, she pounced on him immediately, and it was difficult to avoid being clawed or bitten. Choo-Choo was extremely jealous of the margay and growled fiercely if any of us looked as though we intended to take him away. She had a favourite place on one of the shelves about 6 feet above the ground and as soon as she found the cat would take his whole head gently in her jaws and carry him up to it. Choo-Choo was quite content to lie there all day as long as her companion, who barely weighed 3 pounds, was with her. Margay always made the first move, and when he got up to go out into the garden, only then and not before, would Choo-Choo leave the storeroom.

Choo-Choo was shy and suspicious of strangers, a trait which gave rise to one rather regrettable incident. We had been cleaning the store room, and as a result she and the margay had temporarily moved their den into Jimmy Brown's bathroom. Jimmy was away at the time and, knowing that the store room would be available again in a few hours, I decided it was easier to let them stay there for the time being. Unfortunately, Jimmy had forgotten an essential item of toilet equipment and sent one of his men back for it. Quite without my knowledge the Indian, in a hurry, ran upstairs and blundered noisily into the bathroom. He was lucky not to have been attacked, for Choo-Choo was lying on a shelf just inside the doorway. Instead she took fright and leaped blindly into a nearby window, shattering the pane and falling 20 feet to the lawn. She was quite unhurt and bounced to her feet immediately, running straight for the store room. It was only after I brought her the margay that she eventually calmed down again.

For a jaguar she had the most incredible eating habits. When we gave her a rubber ball to play with, she ate it, so instead I gave her a round can of fruit which served the same purpose but withstood her teeth somewhat better. She also ate my rubber bath mat, part of Jimmy's hammock and a sizable portion of his rubber boots. Nothing of a chewable texture was safe, and she actually swallowed the stuff because it invariably showed in her stool. Like all cats she had a tendency to bury food that was not immediately required. It was amazing how many objects disappeared without trace, only to be recovered perhaps months later when one of the boys was forking over the garden.

Chico, of course, stood out as the big threat to life and limb. Like Choo-Choo he was tolerant of those he knew, to a point, but remained completely unpredictable. He weighed in the region of 100 pounds and was quite capable of killing someone if he wanted to. The time was rapidly drawing near for Chico to remain permanently inside his cage except when under the strictest supervision.

During this period I often kept him on a long lead, made of flexible steel wire attached to a leather collar. He could either be taken for a walk like a dog, or anchored to the base of a tree. Chico was so immensely strong that holding him back when he wanted to go was a bit of a problem as he drew enormous traction from his claws in the turf. He used to love to run down to the lagoon and play in the water. There was never the slightest problem in getting him down there, the distance was only a matter of 70 yards and as soon as I released the end of his lead, he would be off at a gallop. It was all I could do to sprint breathlessly along at the rear with two hands very ineffectively holding the 'reins'. When he got to the water's edge, he romped in with a splash and then sat there with only the top of his head above the surface. All jaguars are excellent swimmers and Chico was no exception. He would often swim up and down the lake for a quarter of an hour while I held onto his lead and ran along the bank. He was easy to control the moment his feet were off the bottom, but if I ever wanted to bring him home before he was ready, he would dig his paws into the ground and give me a tug of war that I couldn't always be sure of winning. On one such occasion I had, after a considerable struggle, persuaded him to come as far as the garden gate. At this point Chico decided he'd had enough. He rolled over on his side, looked at me and gave a disagreeable snarl. I gave his lead a gentle tug and said, "Come on Chico, another few yards and we're home." As far as Chico was concerned that last tug was as much as he planned to take. He reached with one front paw behind his neck, got his claws under the leather collar and with one jerk pulled it cleanly off his head. Chico stood up, the unshackled Samson of the animal world.

This need not necessarily have been a serious problem. Chico had been free before, and his cage was only a few yards away. I had merely to open the door, toss in a piece of fresh meat and he would have been in like a shot. Unfortunately, at that precise moment a distraction appeared. It was the most important man on the ranch, the loss of whom, if only for a few days of disabled sick leave, would have been a catastrophe. It was the cook. He entered behind my back at the far end of the garden. He was of African heritage, and gifted in the African art of carrying loads on the top of his head. Today he excelled. Both hands were free, swinging loosely at his sides, while 4 feet of neatly chopped and stacked firewood were perched deftly aloft. In his wake followed the indispensable Irene, likewise endowed with a wickerwork basket of fresh vegetables.

Chico saw them first. There was no way of stopping him, he trotted past me in a determined advance. I turned, saw the impending disaster and gave the alarm, "Chico's got away!"

Eyes that had hitherto been absorbed in doing whatever eyes must do in order to achieve such feats of balance suddenly focused on the rapidly accelerating cat.

Irene was the first to react. Despite the abrupt change of pace, the vegetables remained in place for a dozen feet while her ample frame suddenly demonstrated an athletic potential of Olympic proportions. Only a stop watch could have measured the lag between her start and the cook's, when the message reached him a split second later. The kitchen door had not been made to allow two people to go through it together side by side; yet despite the design, accommodate them it did, in a photo finish to the fastest foot race I had ever seen. Only the sight of tumbling firewood and vegetables falling in Chico's path saved the day. He hauled to a momentary stop and sent a bunch of bananas skidding across the yard with a ferocious bat from a front paw. Then back to the chase, but the kitchen door slammed shut only inches in front of his nose, and perhaps somewhat bewildered at the disappearance of his quarry, Chico turned and ambled quietly away to his pen.

Strange as it may seem, Jezebel, the harpy eagle, was treated by everybody with almost an equal amount of respect as Chico. The harpy is the world's largest bird of prey, with the exception of the condor. As the condor is a scavenger and does not actually kill its own meat, the harpy is by far the largest of birds that do. Jezebel stood 37 inches tall and weighed 27 pounds. Her wingspan exceeded 6 feet which was not great for a bird of such weight, but the wings were extremely broad and powerful. The reason for their shortness is that the harpy is only found in dense forest and longer wings would hinder its flight through the trees. The bird's most frightening feature were the enormous talons. Her legs had the thickness of a child's wrists and the claws were a full 6 inches long. The rear talon was the fiercest, curling inwards in a sharp, cruel spur. She was beautiful in plumage with a pure white breast and dark-grey wings, but her face was almost sinister in appearance. It was round like a huge owl's and light grey—an effect that was achieved by a halo-shaped ruff and crest. The eyes were cold, calculating, almost hypnotic. A translucent nictitating membrane moved sideways across them, instead of up and down like ours; an asset which enabled the bird to see perfectly at all times, even while blinking. Her beak was imposing, straight out for 4 inches and then strongly curved down at the tip. However, the harpy rarely uses this as a weapon, and relies entirely on its talons for attack and defence.

Jezebel was the second eagle of that name I had kept. The first was sent to a British zoo the previous year. Both were females, which grow one-third larger than the males. Their favourite prey are monkeys and sloth which they tear from the trees, swooping down at speeds up to 50 miles per hour. Primates are so terrified of the harpy that I have seen them leap 100 feet in sheer desperation to avoid their clutches. They will also kill young deer, and any of the smaller ground mammals they can catch. On one occasion, I was standing quietly under a tree calling aguti, simulating their distress call by blowing through a rolled leaf. All at once I had an uncanny feeling that I was being watched, and turning slowly around, found myself face to face with a large harpy. Her approach had been completely silent, and she was standing on a bough a dozen feet above the ground and not 10 yards from me. Fortunately, she realized I was not an aguti and was content to inspect me curiously, turning her head 180 degrees upside-down, first one way, then the other. The harpy has virtually the same mobility in its neck as an owl, and when inquisitive, gyrates its head in all directions in a most amusing manner.

Jezebel had been captured in a startling way. I was on a bat-collecting expedition, moving along a forest trail with several Indians in single file. Suddenly, there was a shot from the rear of the column and doubling back I found one of the men had brought down a harpy. I was thoroughly annoyed about the senseless 'killing', but all at once the bird moved and began to recover. Swiftly we tied her legs together, and she soon revived completely. None of the shotgun pellets had done any noticeable damage, but merely knocked the eagle unconscious. She was one of the most savage things I ever handled and, lying on her back, struck out with her talons in lightning kicks. We made a palm-leaf basket and strapped her inside it, binding her wings to her body to prevent them being damaged. In this manner, we carried her two days back to Dadanawa.

Although Jezebel was a full-grown mature bird, I was determined to tame her, and started off by fitting her legs with leather jesses and tethering her to a perch. At first, I offered her small pieces of meat on the end of a stick. She refused all for five full days, invariably striking the stick and often grabbing hold and breaking it with her talons. Finally hunger prevailed and she began eating. From that point, progress was slow but steady. Gradually the stick was shortened, and then the big day came when she took a piece of meat, apprehensively, but gently from my fingers. The next phase was to make her come to me. For this I wore a heavy welder's glove on my right hand, under a sleeve made from four thicknesses of tanned deerhide. The sleeve covered my entire arm from the fist to just below the shoulder. So equipped I placed the meat on the gauntlet and offered it 12 inches in front of her. She leaned forward and took it with her beak. At the next feeding I held it two feet from the perch; she tried to reach it but couldn't. That day she did not eat. Four days passed before Jezebel bridged the gap and finally jumped. For the first time in my life, I had 27 pounds of the world's fiercest eagle standing on my gloved fist. Reaching carefully with my left hand, I took the jesses and tucked them into the sleeve, clasping them tightly in my fingers. Now Jezebel had only a few inches of play in the straps and I could prevent her either striking my face or walking up the arm to my unprotected shoulder. She shivered a little, then looked at me and whistled shrilly. Ten more seconds elapsed before she bent her head and quickly devoured the meat. As she finished, I turned around so that she was looking back at the perch. I gave a slight heave; taking the cue she opened her wings and hopped onto it.

The rest of the story was months of repetition. I fed her twice a day, morning and evening, gradually increasing the distance. Within six months of capture Jezebel was flying 200 yards from perch to fist. Every flight was a magnificent experience. She flew with slow but mighty wingbeats, head thrust forward, talons tucked back beneath her tail. Fifteen feet from me she would spread her wings like the flaps on an aeroplane, and roll her legs forward, pointing straight at target; talons extended, head and crest up and alert. Then she sailed in with a thud, and often, with a tail wind, the impact was sufficient to send me back on my heels a pace or two.

Jezebel became trained, but not tame. She came because I had the food; she trusted me and flew to no one else; but I was always cautious. Harpy was a wild jungle bird and never let me forget. She kicked at my face dozens of times but I was lucky. Only once did a talon actually reach and scratch my cheek just below the eye; fortunately at the very end of her lunge. On another occasion she landed on my head, but realized the mistake immediately and took off again to make a second landing in a tree. Apparently I made the blunder of holding the gauntlet too low; for eagles, like trained falcons, tend to land on the highest object.

The nearest I ever came to serious injury took place rather unexpectedly. Jezebel was on the gauntlet feeding, when suddenly a Wapishana ran up with some urgent news about a sick horse. His arrival frightened her, and she tightened her grip defensively. One of the talons pierced the leather and the welder's glove, and cut through the back of my hand so deeply, that the point showed under the skin in my palm. It happened so quickly that all I heard was a crunch and felt only a slight dull pain. Fortunately the talon went between the bones, and my hand was not badly damaged.

Jezebel had a propensity for attacking small children, and as a precaution we moved her perch behind a fenced-in area. Despite this she loved to watch the Indian families walk by on their way to the ranch store; and if there were any toddlers amongst them would often fly off her perch and 'attack'. She was anchored by a strong nylon rope attached to both jesses which I inspected daily for safety. The rope was 6 feet long and jerked her to the ground after the first wingbeat. Jezebel never learned, and the crash landings didn't seem to daunt her spirit. She had an evil eye for the children and given the opportunity would no doubt have tried to kill one.

Her arch misdemeanor luckily did not involve a child. One afternoon I stood back amongst some trees about 100 yards from her perch, and, at a signal, an Indian helper released the nylon rope. I held up the gauntlet and she came flying towards me trailing the jesses. She set herself up for the landing, as usual, but suddenly as she touched the leather, gave another wingbeat and flew over my shoulder. I turned around just in time to see a stray dog, the size of a wire-haired terrier, running through the trees. Jezebel was after it, and with a shout I was after Jezebel. The unfortunate dog made a desperate effort to dodge, but mighty black wings spread over the fleeting prey like an umbrella and there was a strangled yelp. It was impossible to pry her talons free. She stood defiant on its broken back and gave the piercing whistle of victory. In that moment Jezebel proved she was still the ruthless jungle predator we had captured many months before. Nothing could ever change that.

There were already enough animals in the grounds of the Dadanawa Ranch headquarters to stock a reasonable-sized zoo, and new arrivals continued to appear at the gates. Parrots, macaws, toucans, capuchin monkeys, marmosets; and even a pair of baby nut-brown tapirs, beautifully marked in white stripes and spots. However, two animals were destined to arrive that would be of major significance. A male ocelot and a full-grown wild puma.

Once a week I patrolled the whole ranch in the aircraft, landing at each of the seven outstations. I usually took off from Dadanawa and headed west to Tar War Wau, Raa Wau, and Tamtoon, our three posts along the Brazilian frontier. From Tamtoon, I flew east to Wari Wau in the south of the ranch at the forest edge, and from there north east to Katalier Wau and Weri More by the Kanaku Mountains. The final hop took me to the northern corner at Atamadpau, from where I doubled back to Dadanawa, which lay somewhat east of ranch centre. The entire flight took from four to five hours depending on how long I conferred with the *capitaso* (foreman) at each stop.

Katalier Wau was the most difficult of the airstrips to get into. The approach was good, across open swampy country with no tall trees; but the strip itself was on the side of a hill that rose unexpectedly out of the flat. As the aircraft approached the threshold it appeared as if it was going to crash into the field instead of land on it. Rather like driving a car along perfectly level desert into the side of a pyramid. To make a safe touchdown it was necessary to flare out in an incredibly nose-high altitude. This meant coming in faster than normal to prevent stalling out. Once the wheels were on the ground I could cut the throttle and the steepness of the hill was more than enough to stop the aircraft rolling without use of brakes. Take-offs were marvellous—regardless of the wind direction I just flew off downhill.

Early in November 1965 I flew into Katalier Wau on one of my routine missions. The *capitaso* met me at the strip and together we walked to the house a few hundred feet distant. As we approached I was surprised to see a little Indian girl sitting on the bare sand in front of the bunkhouse playing with a large ocelot. I walked over and bent down to stroke it. As I did so, the animal bounced to its feet and dealt me a wicked scratch on the forearm and I was lucky to get out of the way before it bit me. It stood there, snarling like a cat straight from the jungle, yet the little girl put her arms affectionately around its neck and pulled it towards her.

"*Auna y naidfan donnyor* [he doesn't like men]!" the *capitaso* said, with a laugh. "*Mara y'pai wrunnow ramtan* [but any female can handle him]. *Kaimen waichpan peu turryan* [why don't you buy him]?"

By this time the ocelot, which I noted was a male, had pressed itself firmly into the girl's bosom, with both front paws on her shoulders. It continued to growl, and was careful to keep its head turned in my direction. According to the *capitaso*, the animal could only be handled by females, and yet he was trying to persuade me to buy it!

As I weighed up the situation, the little girl got to her feet. She took hold of the cat by one front paw and dragged it crudely, yet lovingly, across the yard into the house. Apparently, it had been originally captured by a balata-bleeder in the forest, and spent several months at the latex camp, where it was raised by the man's wife. Recently, at the end of the bleeding season, the Indians had returned to the savannah and sold the ocelet to my *capitaso* for two baskets of farine.

Admittedly, it seemed a pity to take a pet away from the girl that she was so obviously fond of. On the other hand, past experience proved that wild animals never survived for more than a few months in an Indian household. They contracted internal parasites and chigoes, and if those didn't kill them, malnutrition or feline enteritis did. A growing ocelot requires a substantial daily diet of raw meat to keep it in condition, and the Indian family cannot stand the added burden of a ravenous carnivore. When other problems, such as infectious bowel disorders set in, the animal must receive prompt antibiotic treatment or it will die.

So, despite the temperament, I bought the cat, hoping that Irene could handle him and would welcome a pet that only she was the master of. We had no cage or wooden crate, so we persuaded the little girl to put him inside a sturdy cardboard box found in the bunkhouse. This was tied securely shut with strips of rawhide and numerous air holes were punched in all sides. The box was placed on the rear seat of the aircraft, held in place by a safety belt.

After completing the warm-up and cockpit check I looked round to make sure that the 'passenger' was all right. No movement showed in the box, and I pushed forward the throttle. I held her on the ground until the very end of the strip, for despite the downhill run it was still very short. Then, with 70 miles an hour on the clock, pulled her off.

Zero-Two-Zulu had just broken ground when suddenly there was a blow on my shoulder and the ocelot leaped across and landed on the instrument panel. It turned and spat, then flicked a forepaw viciously at my face. Instinctively I lurched backwards, at the same time inadvertently pulling on the stick. The plane zoomed up in a climb and the stall hooter began to blare. Frantically I thrust the controls forward and for precious seconds we hovered between flying and falling. Gradually the air speed crept back to 70 and I held her in a shallow climb.

There was barely room for the animal to stand on the panel so it slumped into a lying position with one paw hanging over my altimeter. As I moved the stick to steady the aircraft it struck at me again, and once more I had difficulty in holding the nose down. After levelling her out I reached down cautiously between my knees and readjusted the seat as far back from the panel as I dared, while still maintaining rudder control with my feet. I toyed with the idea of circling and returning to the field, but the landing at Katalier Wau was difficult enough under normal circumstances, without trying with an ocelot as co-pilot. Dadanawa was only 30 miles away, just fifteen minutes flying, and the strip was longer. I leaned Zero-Two-Zulu into a shallow turn and set course for home.

Unfortunately it was a typical hot afternoon with surface temperatures of 95 degrees and the sky filled with towering cumulus clouds. Ideal conditions for turbulent flying. Every time we hit a patch of rough air and the airplane rose or fell 50 feet, the ocelot showed its resentment by taking another swat at me. This raised the ominous question, how would I be able to land? At lower altitudes turbulence would increase. On the final approach to the Dadanawa field I would be down to 100 feet and coming in over open plain, with a transition to woodland and then water as I crossed the Rupununi River. These three surfaces each reflected heat in a different manner. Experience warned me that the last half mile to the runway was always extremely bumpy. I couldn't afford the risk of an angry cat clawing at my face while flying so near the ground, at only 5 miles an hour above stall. Furthermore with the cat lying on top of the panel my forward vision was seriously impaired. There was no alternative—the cat had to move!

I waited until we were overhead the ranch house, then put Zero in a steep climb. The ocelot dug it claws into the panel and hung on. At two-five I levelled out and waited for the cat to relax. A few minutes later it retracted its claws and shifted a little. As it did I pulled back hard on the stick and stood the plane on her tail. She hovered for a moment, then stalled. The nose pitched forward like a stone. I fire-walled the throttle and let her dive. The airspeed indicator hit the yellow at 140 and I pulled out. The cat had one paw around the fuselage brace and its teeth sunk into the edge of the panel. I tried a 70-degree bank to the right, and then one to the left; I rocked the wings up and down like a see-saw, all to no avail, the cat held on. In desperation, I climbed once more. Two thousand, two-five, three; as the needle touched three-five I cut the power and eased back on the yoke. The stall warning began to blare. I gave the stick one short tug and suddenly kicked full right rudder. The nose and right wing fell together and we were in a spin, heading earthwards with the whole world revolving like a mad cartwheel. I had a vague recollection of a striped blur flashing past me. I neutralized the controls and gave full opposite rudder. The G-forces tightened the skin at the corners of my mouth, would it never stop spinning? Was I applying the right correction? We were rotating to the right, so the answer must be left rudder! Then, gradually we stopped, and with a sigh of relief I applied the power and levelled out.

The ocelot was gone, and the entire cockpit seemed to have been sprayed with a black liquid the origin of which was odoriferously obvious. I looked around and there he was, behind the back seat in the baggage compartment. Only his head and two front paws peered over the top. Cautiously, I began to lose altitude, and gently nursed Zero-Two-Zulu on to the approach. At any moment I expected the cat to wreak the vengeance he deserved by sinking ivory fangs into my sweat-soaked neck. I let down the flaps and we sailed in over the trees and across the river. The wheels bumped on the grassy turf and I tensed myself for the attack. But it never came, we were safely home.

After such an upsetting experience I anticipated trouble in getting the animal out of the aircraft. I taxied up to the house, switched off, and left him in the cabin while I spoke to Irene. Fifteen minutes later she was cradling the cat in her arms. He purred contentedly and sucked affectionately on her forefinger. "Shadow wants a bottle," she said kindly, and carried him off into the kitchen. Our newest pet already had a name, and another friend.

Shortly before Christmas a cat began killing livestock within a 5-mile radius of the houses. Most of its prey were small calves, sheep and pigs and the fang marks on the throats of each victim that we found pointed conclusively to the work of a puma.

It had been several years since one had taken to marauding so close to the headquarters. Examination of the spoor indicated the culprit was a male. Here was a combination I had sought for some time, a male puma on my doorstep. Capture him alive and I had a mate for Leemo!

For a year or so I had experimented with a $C0_2$-powered tranquilizing gun, which fired a syringe in the shape of a dart. The syringe contained a drug called Anectine, and as it struck the animal and the needle entered the skin, the impact caused a light .22 powder charge to fire the plunger forward. Injection was instantaneous. Maximum range of the gun was only 40 yards, but it had proved extremely effective on wild cattle, living in thick forest, that could be stalked to that distance fairly easily. Effects of the drug varied considerably, but some animals had fallen soundly asleep within five minutes and remained so for an hour or more. Without a doubt this was the most practical way of capturing my puma, and plans were laid accordingly.

Men were sent out to patrol the range daily, and after ten days of hunting a fresh kill was found in a swamp, only 4 miles from home. I was soon on the scene in the Land-Rover, with three Indians, the tranquilizing gun, Manager and four other hounds.

The prey was a young calf and only a small part of a forequarter had been eaten. It had probably been killed shortly before dawn, and the puma, with daylight approaching, had left it after a hasty meal; no doubt with intentions of returning the next night.

It was eleven o'clock, but, even with the trail at least six hours old, Mr. Manager strained excitedly at the leash when we showed him the kill. Without further ado we turned him loose.

In a matter of minutes Manager was away through the long grass bawling at every step, and the other hounds were released after him. The trail took us along the edge of the swamp and over the Dobar Wau, a narrow tributary of the Rupununi. Across the stream the pack broke into an excited chorus; the cat was on the run. It was a short chase and after 200 yards he treed in a curatella tree, not 20 feet above the ground.

He was a fine young animal with rich red coat and lay wedged in a fork with a cheek pressed hard against the trunk. Slanted eyes looked down disdainfully at the leaping dogs, and white lips twitched in an occasional snarl.

The three Indian trackers watched rather incredulously as I opened a small metal box, selected a loaded syringe, and fitted it carefully into the $C0_2$ gun's chamber. I took careful aim and fired. There was a hollow thump as the gas exploded, and the aluminium dart hit the cat in the hindquarters. He gave a vicious spit, clawed out the syringe and leaped nimbly to a higher branch where he stood balanced on four feet. The pack became frantic and Manager sprang at the tree and scrambled up to the first bough where he barked furiously. I picked up the fallen dart and opened it to find that the drug had been successfully discharged. It was now just a question of time.

After four minutes the cat began to sway and I ordered two of the men to leash the dogs and take them away to a safe distance.

The puma looked completely bewildered and pawed at his face in a desperate attempt to keep awake. Suddenly he tottered, made one last vain attempt to regain balance, and fell out of the tree sound asleep.

We tied the fore and hind paws together and cut a pole which we passed between them. The cat weighed about 100 pounds and two of us were able to carry it easily while the somewhat mystified dogs followed at a distance.

The Dobar Wau was running knee deep and I chose a firm sand bar as our crossing point. I had taken about two steps into the water when there was a wriggle beneath my foot and a sharp pain struck the upper part of my instep like the puncture from a rusty nail. It was barely ten yards to the other side, yet by the time I got there a column of dull pain had already travelled up my leg to the groin. A small hole oozed blood on my bare foot.

"*Uburr* [sting ray]!" exclaimed the tracker ominously.

Next to snake bite, or an amputation of one's private parts by a piranha, a lash from the tail of a sting ray is about the most incapacitating thing that can happen to anyone in the wilderness of South America. There is a sharp serrated barb an inch or two below the tip of the fish's tail which is covered in a poisonous grease. The sting ray likes to bury itself in the sand to rest, and if a person inadvertently steps on its back the tail whips over and the barb is driven into the leg. Such wounds invariably turn septic and if untreated cause widespread sloughing of flesh around the site and can take six months to heal.

Fortunately I had dealt with a case the previous year with moderate success, and later discussed first-aid procedures with a missionary doctor at the Lethem frontier post. The essential tools were in an emergency medical kit in the Land-Rover and one of the Indians ran the short distance to fetch them.

I took a new surgical blade and made an X-shaped incision, each leg of the cross about half an inch long. These were cut to the full depth of the sting, some three-quarters of an inch, and allowed to bleed profusely. While one of the Indians held the wound open between his two thumbs, I filled a medical syringe from a 100 cc bottle of full strength hydrogen peroxide. Thrusting the needle as far into the incisions as it would go, I injected the contents. The action of the chemical produced a bubbling white foam with extremely strong cleansing properties. After fifteen minutes I had used the entire contents of the bottle and the wound was perfectly clean. I was amazed to discover that the pain had largely disappeared and had no difficulty walking.

This interruption had seriously hampered our progress with the puma. He was still asleep, but it would take us a full half hour to reach the ranch and there was no more time to waste. The Land-Rover had a closed cab and open tray like a small pickup truck. It was impossible to think of loading the cat in the back with the dogs, so we removed the pole and put him in the front seat, with an Indian holding him steady while I drove.

Three miles from home the cat began to get restless. His head lolled from side to side and the eyes were open. I waved my hand across his face but there was no reaction. My foot pressed further down on the accelerator and I began to dodge the bumps and ruts as hard and fast as I possibly could.

Two more hills, one small creek, and home was in sight.

"*Tarro y pokodan* [he's waking up]," the Wapishana warned. With my peripheral vision I noticed the cat's jaws begin to move and the eyes seemed to register now what they were seeing. Cautiously the Indian changed his grip on the loose skin of the animal's neck and moved as far away as possible. Suddenly the puma opened its mouth and screamed. The Indian let go and hurled himself from the car and the cat fell towards me. I pushed at him with a hand and caught him in the shoulder. He tried to bite and swung his tethered front paws in my direction. The wicked claws were out, reaching for my shirt. It was time to leave. I opened the door and jumped clear. The car swung off the trail, bounced over rough ground for 50 yards and stalled, hurling the two dog handlers to the floor of the tray in a heap amongst their hysterical charges. Two of the dogs got free, leaped off the back and charged through the open front door. In an instant there was a bedlam of barking, yelping, fighting dogs, and snarling, spitting, clawing, biting puma. It was almost more than the combined efforts of two of us could do to pull them apart, while the other Indians restrained the rest of the pack and watched with visible concern and shrouded mirth.

We were less than a mile from home so with the help of a lasso and another pole we were able to carry our captive the rest of the way. At least he had recovered from the drug, for even with the most carefully measured and administered doses this is never a certainty. As soon as we arrived I treated my foot with another bottle of hydrogen peroxide and bandaged it up. There was some swelling but only slight discomfort. It healed miraculously with no after effects.

A cage was built for the puma adjoining Leemo's, so they could see and sniff each other through the wire-mesh. It would be quite some time before we dared to put them actually together. The male settled down almost immediately and began feeding off raw meat the next afternoon. It seemed a pity to cage such a magnificent animal, but unfortunately his taste for calves was not compatible with ranching objectives.

We called Leemo's expected mate Roland.

XIV

MORE ROUND-UP, AND THE WEEK THE ARMY CAME

The new year round-up season started on 5th January. As customary, it began with the section immediately surrounding the headquarters. This involved driving large herds across the Rupununi River, so the early months of the year, with predictably low water, was the most suitable time.

Cang had long since gone to graze in greener pastures, but Matador and Guitar were still solid horses, despite their advancing years. Matador was the ruggedest pony I ever put a saddle on.

He could run all day, and over a long stretch was the fastest horse on the ranch. It didn't seem to matter what the terrain looked like—rough boulders, sharp quartz, soft mud or ant hills. He could take them all in his stride and rarely stumbled. Because of his incredible surefootedness, Matador became my favourite roping horse, particularly when we were after the swift and wily bush cattle. He could hold back the strongest bull, and when I dismounted to tie it up, could be depended upon to stand fast and keep the rope taut.

First round-ups in any year were always slow in attracting vaqueros. Many of them had worked until late in December and, with the money they had earned, purchased all their current requirements. An Indian who has his quota of cloth, ammunition, cooking pots and supplies needs no work. He's content to sit at home, until once again hunger, or necessity, drives him to the abominable status of employee. Other factors were the Christmas and New Year celebrations. The Wapishana know little or nothing about the religious significance of the former, and have no traditional calendar which can pin-point the latter. Christmas was imported to the savannahs about fifty-odd years ago. It came in various forms. The religious rendition was expounded by a few zealous Roman Catholic Missionaries who suffered almost unbelievable privations, trudging for miles through forest and across fiercely hot, dry or flooded plains, between the scattered villages. They were completely at the mercy of the Indian's good will to provide them with food and shelter. They were like Livingstone in Africa. Like Livingstone, they got malaria, and like Livingstone, many of them died. Unfortunately for them their Christmas message conflicted with an entirely different philosophy widely practised by every hard-drinking settler, gold prospector and travelling adventurer in the territory. It would have been better for the welfare of the Indians if the Christian approach to that last week in December had been the one they adopted. Regrettably, the reverse was true. By the fourteenth of the month, there was hardly a vaquero left on the ranch to milk a cow or tend to a sick horse. Hordes of women were squeezing cashew fruit, pulverizing sugar cane, whipping up bananas and grinding cassava. The juice was fermenting, not by the bottle, or even by the bucket, but in dugout canoes, dragged up on the beach and filled to the gunwales. When the resources of one village were depleted, many of the tribe would move on to the next. In a good year an almost inexhaustible reservoir could prolong a drunken orgy day after day for almost three weeks, and it was unusual to see any sign of the end before 3rd January.

Therefore, 5th January as a round-up day was sometimes not very realistic, and we were invariably shorthanded for the first week. Such was the case in 1966, and, in order to occupy what men I had available, I organized some wild cattle-roping sorties. These were not round-ups of a major scale and any number of vaqueros could participate. We merely flushed renegade steers and bulls out of patches of forest, lassoed them and took them to the corral. When sufficient men finally arrived to launch a giant sweep across the range, the wild animals would be released amongst tame ones and driven off to the trail.

Four bulls had been seen entering a patch of dense cover isolated 100 yards from the heavily forested Rupununi River bank. I sent six men inside on foot, armed with empty 4-gallon tins, while ten of us surrounded the patch on horseback. Cattle are crafty beasts and they were almost certain to break out on the side facing the river, knowing that with luck they could sprint across the 100-yard gap before we had a chance to lasso them. Bearing this in mind, I hid with Matador behind a tree about 70 yards from the Rupununi.

Soon the boys began to create a fearful din with the empty cans, and within a few minutes, two of the bulls darted out and came obliquely towards me. I waited until the leader was opposite my position and gave Matador the spur.

In 50 yards I was alongside his right flank, but with great cunning he turned to the left and directly away from me towards a fringe of curatellas. I followed on his heels but was unable to make an early throw because of the trees. Soon only a few yards remained before he would make the dense forest, but there was a small gap ahead. One swing was all I had, and a throw straight past my horse's head. The heavy rope took him perfectly by the horns, but suddenly he swerved hard to the left. Desperately I tried to turn Matador after him and at the same time pick up the slack of the rope which was paying out faster than I could handle under the horse's neck. The turn was too tight, and there were too many trees. I had a 40-mile-an-hour horse going one way, 1,200 pounds of wild bull going the other, and 90 feet of rawhide lariat connecting the two. The rope was anchored under my thigh on the right side of the saddle and now passed across my leg, under Matador's neck and off to his left quarter. Disaster came in a second. As the rope jerked tight, it picked up horse and rider and threw us in a wild somersault to the ground. I never left the saddle, because I was tied to it; I felt a bone smash in my shoulder and then we were being dragged across the ground with Matador thrashing frantically to regain his feet. Suddenly the bull stopped, turned and charged. I spurred Matador in the flank with my free leg but he didn't get up. We lay there defenceless and waited to be killed.

Two vaqueros galloped from the trees and cut across the bull's path, screaming wildly and throwing their hats in its face. The Longhorn skidded almost to a halt, hooked at the hats with his horns and then charged the nearest cowboy's horse. The Indian gave whip and spur in earnest and was off with the bull hard on his tail. I braced myself for the shock as he reached the end of the rope, but it must have been close to breaking from the earlier jerk, and parted easily. Breathlessly I watched bull and rider disappear into the trees.

Another vaquero rode up and two of them pulled me out from under the horse. Matador got to his feet and shook himself. He appeared to have survived the ordeal with nothing more than a rope burn under his neck. I had not been so lucky. There was a severe pain from the base of my neck outwards to the left shoulder. The clavicle had been forced upwards and out of place a full 2 inches, and when I moved the arm there was a feeling of grinding bone at the joint.

"*Peu tawut sootan* [your shoulder's dislocated]. *Kaimen waichan wom tootan* [let's jerk it back in place]!" exclaimed one of the vaqueros gravely. He took hold of my arm, moved it first forward, then backwards, then gave it a smart jerk outwards. I almost fainted with pain and nothing slammed back into place.

"I think you'd better leave it alone," I said, "I'll ride back to the house."

I scrambled aboard Matador and walked him gently home. It seemed fairly obvious that I was suffering from more than just a dislocated shoulder and would need qualified medical help. Georgetown was the nearest place I could find that.

Irene packed me an overnight bag and I got in the 'Grey Ghost' and drove the 62 miles to Lethem, arriving late that night. I slung my hammock in a spare room at the saloon and next morning was fortunate to get a ride on a chartered DC3 that flew in with freight.

Four hours later I was under an X-ray at a Georgetown hospital. There was no dislocation, but I had a fracture of the scapula, which in addition had forced the collar bone upwards. There would be no need to set it, the doctors informed me. The break in the shoulder blade was in the form of a puncture; probably caused by a stump being driven through it in the fall. It was only necessary to strap my arm across my chest and leave it for six weeks.

Three days later I was back at Dadanawa on the round-up. I'd learned a lesson about lassoing wild bulls through thick trees, and was branded with one shoulder higher than the other to prove it.

The country's independence from colonial rule was scheduled for 26th May 1966. The last few years had seen some turbulent times in the city and coastal belt, although none of these disturbances had ever affected the far interior. British troops were stationed in the troubled areas and occasionally platoons were sent to the Rupununi District on a scheme called 'adventure training'. Such excursions were, in fact, a thinly disguised excuse to get guard-duty-bored soldiers out into the fresh air for rest and relaxation. Discipline on these trips was maintained at a minimum, and therefore I think it proper to omit the regimental name of the group that came to Dadanawa.

I received a signal from the Georgetown office to expect twelve soldiers under the leadership of a Major Alistair Montagu to arrive at Wichabai on the next charter. The plane was scheduled for Friday noon.

The Land-Rover was hardly adequate transportation for a platoon encumbered with hobnail boots, back packs, and rifles. So I drove over to Wichabai in the tractor, pulling a large trailer.

The war surplus bomber, exactly one hour late, taxied towards me and swung its tail to within 10 feet of the parked tractor. A bronzed pilot hiding behind dark glasses leaned nonchalantly out of the window. He smiled mischievously as the prop blast blew a storm of grass and sand over me and the vehicle, and then cut the engines. Big double cargo doors yawned open, and the first soldiers took the 5-foot jump to the ground throwing their kitbags ahead of them. A crew member fitted the metal ladder in position and the officer in charge, complete with swagger stick, descended.

I walked over and extended my hand.

"Major Montagu, I'm Stan Brock. How d'you do."

"He's no major," interrupted a pimply-faced private. "He's a flippin' second lieutenant."

Lieutenant Alistair Montague flushed slightly and tapped his stick in the palm of his hand.

"How d'you do, Mr. Brock?" he said, in perfectly cultured tones. "Jolly nice of you to invite us here." He paused for a second, looking at his troops. Some had already wandered off into the savannah, while others were squatting on their kitbags having a smoke. Lieutenant Montagu raised his voice, "Now come on, fall in you lot, there's good chaps."

The soldiers casually arranged themselves in two ranks, and were asked to extinguish their cigarettes. They had one or two things in common. Their faces were red from the sun, and suffered either from shaving rash or pimples or a combination of both. The youngest was about 18 and the eldest no more than 21. There were fat ones, thin ones, tall ones and short ones, and all wore crumpled khaki denims.

I eyed them for a moment, and the thought suddenly crossed my mind that these were probably the first teenage boys from England to stand on the Wichabai airstrip since I first stood there myself thirteen years before.

Lieutenant Montagu took a note pad from his top pocket, and read the roll call.

"Corporal White?"

"Sir!"

"Rifleman Bonny?"

"Sir!"

"McLeod?"

"Sir!"

"Pulvertaft?"

"Sir!"

Twelve men answered in order, apparently no one had been lost *en route*.

"Now men, this is Mr. Brock. . . ."

"I like 'is bare feet!" one interrupted.

"Mr. Brock is in charge from now on, and I want you to show him the same respect that you show me. . . ."

"Cor! This is goin' to be a right lark," said another and his remark was drowned in laughter.

"Where are your rifles?" I enquired.

"Oh, we didn't bring those," the lieutenant replied, then, after a pause. "I say, you're not expecting any trouble are you?"

We arrived at Dadanawa for a late lunch. Chico, Choo-Choo, and Leemo had been locked up, for obvious reasons, and everyone was warned about sticking their fingers through the wire-mesh, or mistaking Jezebel for a friendly parrot. The men were billeted in the vaqueros bunkhouse, but meals were served around the long table in my dining-room.

Lieutenant Montagu questioned me about activities that would keep the men occupied.

"Got to keep them busy, you know," he emphasized, gripping his pipe bowl. "They're not a bad lot, but with nothing to do they'll soon get up to mischief. . . . Another thing," he warned, "sometimes they say things they don't mean. Damn embarrassing yesterday when we were leaving the barracks, a captain from the Grenadier Guards walked by and one of them said in a loud voice, 'Cor, look at that toffee-nosed bastard!' "

I paused for a moment, wondering what Charlie might have done, if he'd suddenly been descended upon by a platoon of soldiers. Probably take them out in the bush somewhere, and hope they'd get lost.

"Well, we're cutting a trail in the forest about fifty miles from here," I offered. "Between Maroodi Mountain and the Kassi-kaityu. Maybe they could work on that?"

"Oh rather!" the lieutenant replied enthusiastically. "What a jolly good idea."

"It's a long way though, through little-known jungle," I cautioned.

"That's all right," he answered. "Make it a better exercise."

"Only thing is you might lose a few of them," I suggested. "Does that matter?"

"Good Lord, yes," he retorted quickly, with eyebrows raised in a look of deep concern. He removed his pipe and craned his neck closely towards me. "It's their mums, you know. Ask all sorts of difficult questions—go to their M.P.—comes up in the House and then there's a frightful stink."

"Well, that kills that one," I replied. "Maybe we should just go out near here and hike about the country a bit?"

"Damn good idea!" exclaimed Lieutenant Montagu.

Next morning, after breakfast, we left on foot for the Kanaku foothills. They lay about 3 miles away and the terrain was interesting enough to occupy everyone for the entire day. Field rations were stuffed into back packs, and Rifleman Bonny was given the additional burden of an enamel pot.

Corporal White lined up the troops and they set off in a rather straggly formation with Lieutenant Montagu and me bringing up in the rear.

"Not a bad sort, Corporal White," the lieutenant remarked. "Used to be in the Irish Guards. Got kicked out for swatting a naval officer in a London pub. . . . Doesn't know we know, of course. . . . That was the last straw, he'd been on a few other charges before . . . arrested on his wedding night for an indiscretion in a fish and chip shop. . . . They let him get married, then drove him away from the church in a Black Maria. . . . After he got out, he changed his name and joined up in our lot. . . . Damn good soldier. . . . Used to be a sergeant—got reduced last year, turned out for an emergency without a bolt in his rifle."

By eleven o'clock the chorus of bar-room songs, lustily led by Corporal White had petered out. Sweat darkened their khaki denims and water bottles were dry. It was a cloudless day, and the sun beat mercilessly down through stagnant air.

"It was cooler than this in Aden!" the lieutenant breathed. "At least you could jump in the canal."

"There's a creek not far ahead, would you like to stop there and rest?" I queried.

"Bloody good idea," he replied. "We can break out the lunches."

Fifteen minutes later we arrived at the banks of a stream, but there was nothing in it but sand and rocks. We walked up the dry bed for awhile and eventually found a single waterhole. It was covered in green slime, and dollops of waste matter bobbed on the surface.

Rifleman Bonny, cup in hand looked at the sight in horror, and blurted, "Blimey, look at the cow. . . ."

"We can all see it," interrupted the lieutenant.

"And smell it!" added a rifleman.

I stepped down and skimmed the scum off the surface with the back of my hand. The sand glittered through the clear water beneath. Rifleman Bonny stared in absolute horror as I took his cup and drank deeply.

"I say, isn't that stuff poisonous?" asked Lieutenant Montagu.

"Not at all," I replied. "Dadanawa cattle are the healthiest in the world."

Someone produced a bottle of pills labelled 'water purifier', dropped a couple in the enamel pot, and filled it carefully with a cup. They watched the tablets slowly dissolve. Corporal White took a cautious sip. He licked his lips, gave a shrug of his shoulders, then announced, "taste's okay."

At once the other men joined in and soon the pot was refilled again and more tablets added.

Lieutenant Montagu sat on a nearby boulder. He hadn't brought a water bottle and was obviously incredibly thirsty. He looked at the men drinking, and then at the cow turds, he sucked at the inside of his cheeks, but there was no saliva. Perspiration dripped from the end of his nose.

Corporal White downed his third mug of water and looked over at him. "Sir," he said, "you look as though you've 'ad it!"

"I'll take a small drink of water if you don't mind," he replied.

The corporal handed it to him and the lieutenant filled his mouth. He avoided swallowing any, and after holding it for a minute, let it trickle slowly through his lips to the ground.

"Just don't think about where it came from," I suggested, "and it'll taste fine. You remind me of the police. They came to Dadanawa the other day, just dying of thirst. Had a problem with their jeep and couldn't find any water. They wanted a bottle of rum and I told them the only stuff I had was contraband from Brazil. The sergeant said, "I don't give a damn where it came from, as long as it's drinkable!"

The next mouthful Alistair Montagu swallowed.

The following morning there was little enthusiasm when I suggested at breakfast that we go for another hike.

"Perhaps we could do something more restful," said the lieutenant. "We were on manœuvres last week, and I think I've done enough foot slogging."

"You've done enough?" exclaimed the corporal in disbelief. "We were doing the foot slogging, you were in that trench at H.Q." Corporal White leaned across the table towards me and waved his spoon at the lieutenant. "He nearly shot down a bloody B29 with a smoke bomb; and when C Company was attacking us, 'im an' the colonel were sittin' on the top of a truck with a pair of binoculars, watchin' ruddy birds!"

"Corporal White," Lieutenant Montagu corrected, "it was a very rare blue-crowned motmot!"

While listening to Corporal White I became aware that Rifleman Bonny was having some kind of a problem. A wet patch had appeared on his shirt, and water seemed to be dripping from the ceiling into his boiled egg. He very politely said nothing, but moved the egg aside, and quickly pushed his teacup into place to catch the drips.

"Well, Father McKenna may come here this morning," I said. "He's a Catholic Missionary with the Wapishanas . . . said he might pop in around eleven o'clock and have mass in the bunk-house."

"Capital idea," Montagu retorted. "But what'll we have if he doesn't show up?"

"Gin and tonics," said a rifleman from the far end of the table.

"Aren't there any girls around 'ere?" questioned another.

"Say, did I tell you fellas about the one I met in Port Mourant last week," Rifleman McLeod interrupted. "She said it was five dollars, and when I said that was too expensive, she said 'er old lady was only one!"

Lieutenant Montagu looked at him with an air of disgust, while the dining-room shook with laughter. "McLeod, we're not interested in your love life," he said. "And besides the medical officer asked you to lay off that sort of thing for a while because he's running out of penicillin."

"Well, let's move out of here anyway," I suggested, "and give Irene a chance to clear the dishes—I'll see you out in the yard."

I went upstairs to get my hat, and suddenly remembered Poppy. She hadn't been feeling too well for a day or so and I'd brought her down from the kennel. She was lying in the centre of the room with her chin on her paws and looked sheepishly up as I climbed the stairs. The front door was closed, and in the middle of the floor there was a gradually dwindling puddle.

Outside, the soldiers were highly amused by the antics of the two monkeys in their pen. Tiny was lying on his side against the wire, sound asleep, and Big and Little Jacko were reaching through the mesh, picking out particles of dry salt from between his hair and eating them.

"How extraordinary," the lieutenant commented.

"Tiny goes there every day for that service," I remarked. "And he's not the only one. We have an old ewe that stands by the cage and lets Jacko milk her. The way her tail wags, I think she enjoys it. Then, there's a hen that lays an egg in the grass there almost every morning. Those monkeys are well fed before we ever go in there with the bananas! . . . Now, what are we going to do? I suggest we take a ride in the jeep and do some bird watching. Any of the men who want to can stay behind and fish or ride. We've got plenty of tackle and horses."

"Excellent idea," agreed the lieutenant. "Let's do that."

"Maybe you'll find another blue-headed what-not!" said Corporal White.

"A blue-crowned motmot," Montagu corrected. "You may all dismiss."

The four days that the army spent at Dadanawa was probably as entertaining for the Indians as it was for the soldiers. Never before had they seen a group of white men behave with the enthusiasm of happy youngsters. They pushed each other in the river, and threw away one another's caps. They borrowed cowboy hats from the vaqueros and sat precariously in the saddle for endless photographs. They tangled each other up in lassos and spent hours throwing them erratically at any dog, sheep, chicken, horse or person that, through ignorance, passed their way. They bought bows and arrows, woven baskets, bead aprons and matapees. I intervened only when Rifleman McLeod tried to buy an Indian wife. "She says her pack ox goes with her," I told him. "What are you going to do in the British Army with a half-naked Indian and a pack ox?"

Finally, the day came for them to leave. The DC3 landed at Wichabai and the troops were lined up in front of its open doors. Second Lieutenant Montagu, swagger stick in hand, stood smartly before the ranks.

"Platoon . . . 'shun!"

"Platoon . . . three cheers for Mr. Brock. Hip-hip. . . ."

There was total silence except for a slight giggle in the rear rank.

"Now, now, come on chaps," the lieutenant insisted. "Hip-hip. . . ."

"Hooray!" came the response. "Hip-hip . . . Hooray! Hip-hip . . . Hooray!"

As the final cheer still echoed through the hills the army scrambled aboard the aircraft. A family of travelling Indians sat resting beneath a nearby tree. I looked at them and could read the thoughts written in their eyes. 'The white man does indeed belong to a very strange tribe.'

I watched as the old Dakota lumbered away into the clouds high above the Kanaku Mountains. I wondered what carefree escapades would confront its passengers next. For a moment, I almost wished that I was on board.

XV

NEW HORIZONS

Leemo had been coming in to season regularly since her third year. The periods usually occurred every three months and lasted for ten days, with a peak on the fifth and sixth days. They were characterized by a swelling of the organ and prolonged, melancholy, coarse meowing. This wailing was the only disturbing part of her behaviour. Her endurance was phenomenal, and on the fifth and sixth days she could meow almost around the clock without a break. On a still night the noise carried for half a mile and there were a number of justifiable complaints from Dadanawa residents who found it impossible to sleep.

An interesting observation I made during these times was her remarkable placidness. She was less interested in food than normal and far less prone to attacking other animals. The sheep and cattle could pass her fairly closely without the usual fear of instant pursuit. She even resisted chasing horses, for it was a favourite sport to stalk them through the grass and then suddenly spring into a mock attack, even striking one on the nose. When they turned and galloped petrified across the savannah, Leemo would give chase with tail beating in the air, until finally the ponies greater endurance would prevail, and leave her breathless but elated in their wake. Perhaps the horses could smell a difference in her attitude during these seasonal periods, for though she might walk within a few yards of them they would only raise their heads in a casual glance and then continue their grazing.

Oestrus had occurred in September 1966, and so I was prepared for another one in December. Roland had settled down admirably in his new surroundings, and from time to time had been noticed taking interest in Leemo as she passed by. Unfortunately, his attentions were not reciprocated, and although he rubbed his neck affectionately against the wire and stared after her with longing eyes, Leemo kept her nose high in the air.

In November I decided to open a gap in the wall between their two cages. This allowed Roland to commute back and forth during the day while Leemo was out, and leave his various alluring smells in her compound. Beano was, of course, removed and housed temporarily elsewhere. The signs of a strange cat having actually prowled in her domain did not appear to bother Leemo unduly. She smelled his spoor, rolled in it and then deposited her own urine on top. The time had come to actually allow the pair to meet.

Roland pranced towards her enthusiastically and put a paw across her back; but Leemo slid from under it and trotted off to a corner where she sat on her haunches staring outward through the wire. Roland's alacrity was not to be tempered so easily and he followed at once. As he approached Leemo swung about and dealt him a solid blow on the nose with a forepaw. Roland sat back on his tail, but refused to retreat. In answer the female stood on her hind legs and hit him with a swift combination; then slipped quickly from the corner and leaped onto her table. This time Roland did not follow, and finally the pair settled down at a safe distance to watch each other.

Within a week they had reached a mutual understanding about the new territory. Roland was allowed everywhere but on the table which remained Leemo's sacred perch. At mealtimes they took their individual portions of meat and devoured them without argument. I felt sure that when she came into season Roland and Leemo would soon become the best of friends.

Her period began on 14th December. I had been out riding all day and did not return until late in the evening. It was already dark, but when Irene told me that Leemo had started wailing that morning I decided to pay her a visit.

As I approached the cage I could see Leemo's dim form on the table. Roland had apparently returned through the gap in the wall to his own quarters for I could see him lying there in the shadows. I opened the door and walked in. "Leemo," I called. "It's me." She sat up and watched. Still talking to her, I walked across, and patted her between the ears. Suddenly there was a screech like a circular saw running at full speed into a block of hard wood, and 'Leemo' went one way and I went the other. It was, I admit, one of my hastier exits through a doorway, and I slammed it behind me in a manner which could be reasonably described as rude. I took a deep breath and peered through the mesh. Leemo gave a squeak and came cautiously through the gap in the wall and walked up to the wire. She purred affectionately and I reached in and scratched her ear.

"I thought you didn't let him sit on your table?" I asked. She took my finger between her teeth and gave it a friendly nip; she seemed to be thinking, that Roland was not really such a bad cat after all.

The change in Leemo's attitude to the male was indeed remarkable. For the next week they lay together on the sandy floor of the house as if they had been raised together as cubs. No longer did Leemo defend her table top, and Roland was allowed to climb upon it whenever he wished. Even the wailing was reduced and although no intimate relations were actually witnessed, I concluded that they had doubtless taken place. I was confident that in ninety-six days time Leemo would produce her first litter.

For the past ten months my riding had been hampered by a large septic sore on the inside of my right calf, at a point where it came in constant contact with the saddle flap. It began as a small blister the size of a match head, but rapidly developed until it was some 2 inches in diameter and almost as deep as the bone. From that stage, which was achieved in three or four months, it got neither better nor worse. I treated it with every form of antibiotic powder I could find, all to no avail. In desperation I showed it to the Indians, and it was systematically anointed with concoctions from every herb, leaf and root that tribal lore recognized as possessing medicinal qualities. The sore flourished.

One day, when the seepage of puss was particularly ominous, I decided that my only answer was a visit to a Georgetown doctor. I climbed into Zero-Two-Zulu, and flew off.

To my surprise the doctor found nothing unusual about the condition. He gave it an almost cursory glance, and announced I had 'yaws'.

"Commonly known as 'bush yaws' in this country," he went on. "A typical case, ulcerating lesions, could later involve the bone, it's a contagious disease caused by the spirochete Treponema. Almost impossible to cure, I'm afraid, old boy."

"Do you mean this bloody thing's going to eat me alive!" I asked, already breaking out into a cold sweat.

"Oh no, nothing like that," he said. "I can give you all kinds of injections, but they never do any good. Usually healing is spontaneous. After a while it just seems to get better on its own. I'll let you have a bottle of stuff and you can give 2 ccs intramuscular in your bottom every day until it's finished."

After a month my rear end felt like a pin cushion, and the task of personally driving a needle into it became less and less a happy one. At about that time an old negro gold prospector passed by Dadanawa on this way to Brazil. Knowing he had probably spent all his life in the forest, I showed him my sore and asked him if he'd ever had one.

"Oh yes, baas, many times," he replied. "Ah even had one on my face." He stretched the leathery skin on his cheek with gnarled fingers, and, as the flesh lightened, the faint trace of a circular scar revealed itself.

"How did you cure it?" I asked.

"Well, most of them does jus' go away by theyself, but ah had one on my arm for two years," he rolled up his sleeve and pointed to a shiny patch of skin on his elbow that resembled a severe burn. "Ah cured that by wrappin' it up in a bandage for a month, an' stoppin' the air gettin' to it. . . . Can't breathe without air, an' the germs jus' die off!"

After two weeks with the same bandage on my calf I could take it no longer. The smell was so bad I could barely stand being near myself, and the irritation and throbbing pain made sleep at night impossible. On the fifteenth day I peeled the bandage off.

The sight almost made me sick. My yellow wound was crawling with maggots!

For a moment I couldn't understand how they got there. Then it became obvious. While out riding the savannah, or working in the corral, flies had been attracted to the wound. They had flown up my trouser leg and somehow crawled underneath the bandage and laid their eggs. Hatching of the larvae takes place in a few hours and they grow rapidly. Now I had hundreds of them, burrowed deep into my flesh, devouring it voraciously.

Such maggots were common place on any wound amongst livestock. A cut no larger than a dime can be eaten away to the size of a man's fist in the space of one week. Yet they are easily killed with a variety of chemicals and aerosol sprays found on the agricultural market. Isolated as we were at Dadanawa we had none of the sophisticated products, but instead relied on a plain sanitary fluid which we purchased in 4-gallon cans.

While on the verge of applying the disinfectant to my leg, it suddenly occurred to me that if I left the maggots alive for a few more days they might easily eat away all the malignant tissue!

Two days later I poured on the sanitary fluid. It burned like acid, but maggots fell out by the dozen. Three applications, and even those deeply imbedded had been forced to crawl forth and writhed dying on the floor. When the treatment was finished the wound resembled a honeycomb, each channel a passageway for a tunnelling larvae.

The maggots had done their work well. Within a few days, pink tissue began to replace yellow oozing puss. Miraculously the gaping hole gradually closed. In a month my infamous bush yaws was nothing more than a hideous memory.

The first half of February 1967 kept me extremely busy with round-up at Tamtoon outstation. Tamtoon was remotely situated in the south-western corner of the ranch and a good two days ride from Dadanawa, or twenty minutes flying. Cattle in the area were both notoriously scrawny and notoriously wild. As a result, they took twice as long to catch as those anywhere else.

One day we were running a herd of 700 across the Kawara Wau river on our way to the corral when a solitary rider approached from the north. I recognized him as a young Indian that I had left at Dadanawa herding spare horses. His sudden appearance at Tamtoon on a tired pony spelled trouble.

He rode up and removed a sweat-soaked piece of paper from inside his hat. I opened it to find a cryptic message from Irene scrawled in pencil, "Please come if you can." The Indian seemed to know nothing of any bother there, so I concluded it might be a personal matter. I left the vaqueros in charge of the round-up and cantered ahead to the outpost. In less than an hour, Zero-Two-Zulu was taxiing to a familiar parking place on the Dadanawa airfield.

Irene was sitting at the kitchen table. She leaned forward on her elbows, and her hands were clasped in front of her mouth. Her eyes stared emptily at the bare table-cloth

"What's the matter?" I asked.

"Leemo," she said, pausing for a moment. "She's dead."

I swallowed hard; thoughts, reasons, all began to flash across my mind.

"How?" I breathed.

Irene looked up. Her eyes red and puffed, her cheeks wet. She opened her mouth to speak, but the words were stifled in a sob. I waited while she blew her nose.

"She got out of her house in the night. There was a hole in the roof. Shadow and she meet, and they fight, and Leemo dead."

I couldn't believe it, a full-grown puma killed by an ocelot. There was 70 pounds difference in their weights!

Irene pointed towards a large cardboard box in the corner of the kitchen. I walked over, and there inside, snarling up at me was an incredibly scarred and battered Shadow.

"He nearly dead too," said Irene. "But he walk this morning, an' he drinkin' 'is milk."

Leemo had been buried at the end of the garden. An Indian who dug her grave told me later that there were no marks on her body anywhere but on her throat. The story of the battle finally began to make some sense.

Who started it we'll never know, but Shadow was certainly on the losing end. His terrible wounds proved that. It was the old story of the lucky punch. Leemo probably had her opponent lying on his back on the ground between her paws. Then suddenly a last desperate move by Shadow as the puma dropped her guard and he seized her soft fleshy throat in his jaws. Leemo's own fangs were powerless, and she could only strafe the ocelot frantically with her claws. Relentless jaws tightened their deathly grip, the windpipe collapsed, and in a minute Leemo could breathe no more.

Nearly fifteen years had passed, and there were already many young vaqueros working at Dadanawa who had not been born when I first set foot on Rupununi soil. To any Indian under twenty I seemed as permanent a fixture as Jimmy Brown, Irene or the mango trees. The dozens of pet animals, the hunting dogs, my special string of horses, all appeared as confirmation that Brock would sooner or later lend some little nourishment to the impoverished Rupununi dust. But within me the voice of new horizons had begun to call.

I had published a couple of books about the adventures of Leemo and the other animals on the ranch, and these had been noticed by the British Broadcasting Corporation. As a result, they dispatched a film crew to Dadanawa and produced two television episodes about my life there. They filmed everything from chasing cattle to capturing giant water snakes and the programmes were later shown in many countries of the world. I didn't realize it at the time, but film work was shortly to play an important part in my future.

The aeroplane was also a contributing factor to my rapidly developing wanderlust. It had flown me across frontiers and unknown jungles. It made mighty spectacles such as the Essequibo and Kaieteur Falls, plunging five times the height of the Niagara into the Potaro Canyon, part of my everyday life. Suddenly, there was an urge to see other frontiers, other jungles, other falls. In an effort to placate these inclinations I mounted numerous expeditions.

The most enthusiastic was the cutting of a 350-mile trail from the southern edge of the savannah across previously untracked forest to the Suriname-Brazil border. This rough-hewn pathway linked the Rupununi grasslands to the extensive Paroe River savannahs, which were only inhabited by primitive Indians of the Trio tribe, and had never been settled by outsiders. Here lay potentials for as large a cattle operation as the Dadanawa. Unfortunately, my board of directors did not agree with this theory; when I suggested driving a hundred Longhorns along the trail as an experiment, I was politely asked to abandon the project.

I increased my collecting activities for museums and zoos, and was fortunate enough to capture a particularly unusual variety of bat later that year. It was quite a tiny creature, as most hitherto undiscovered species are these days, with a body length of only 2 inches and a wingspan less than 5. The colour was grey-brown, and it had a white dorsal stripe, as well as contrasting white bars that extended from the base of its pointed leaf-shaped nose to the corners of the ears. I sent the specimen to Dr. Randolph Peterson, curator of mammals at the Royal Ontario Museum in Toronto, and the world's leading expert on the Chiroptera of Northern South America. As a result of this, and additional specimens that I was able to send him shortly afterwards, he classified it as a new species. He named it *Vampyressa brockii*, which I thought was the highest honour that any scientist could possibly bestow upon a cowboy.

The bat came from a little-known region at the head of the

Kuitaro River which was full of wildlife. The game appeared to have no inherent fear of man. Jaguars, puma, deer and tapir, all stood in the trail and watched us without the slightest sign of trepidation. We even came across a pair of Giant Brazilian otters walking on a sandbar in a river bed, inspecting a caiman basking in the midday sun.

Among the larger animals, the ones I sought most were the giant armadillo, tapir and anaconda. The armadillo, to many, is a familiar creature, a terrestrial burrowing mammal with a protective carapace of flexible horn, and powerful claws on its front feet that function with a shovel-like action. Common species weigh from 7 to 15 pounds. The giant armadillo is a gargantuan species weighing in excess of 120 pounds with a length of nearly 5 feet. This miniature tank has cutlass-shaped claws 6 inches long and can drill a hole in the hardest ground. One I caught by the tail dragged me along on my belly as if I didn't exist. Only by keeping up the chase until the animal was tired, and rolling him upside down to prevent him from digging, was he finally overpowered.

The tapir is probably the hardest of the South American animals to capture, largely due to its incredible strength, but also because of the environment. They live mainly in dense forest near water, although during dry periods, and in mountainous regions, they can get along well without it. Their ability to run with the speed of a horse and agility of a goat makes pursuit in most instances a losing battle. Persistent hounds with better than average wind can bring them to bay, but then you have cornered a beast with the slashing hooves of a stallion and the biting tushes of a wild boar. My captives were usually young animals, and as tapirs are basically mild-mannered creatures, they became tame and tractable within a few days. On one occasion we managed to drive a three-quarter-grown specimen into a heavy game net, set in a gully at the entrance to a lake. Although we eventually brought the animal under control, it was only after a half-hour battle in waist-deep water, and virtual demolition of a hitherto considered indestructible nylon net.

Anacondas aren't my favourite. I've been bitten and squeezed by a few of them. They're strong, and in a tussle with man have an unfortunate habit of winning. Many horrifying tales have been told of the anaconda. Some are true, but in most the imagination like the length of the snake, has been somewhat stretched. Nevertheless he can be a dangerous customer.

The female gives birth to live young. As many as forty may be born at one time. The species is common, yet rarely seen, for they spend the greater part of life dormant at the bottom of deep pools in quiet, sluggish stretches of river.

Anacondas are gluttons, feeding only on what they kill themselves and, if the prey is large, lying for weeks digesting it. During this state of satiation they are quite harmless, and often so lethargic that at first glance appear to be dead.

Blessed with only slight intelligence, they are incapable of little thought beyond interpreting the hunger transmissions from stomach to brain. After that, grabbing the first thing that passes by is a reflex action, whether it be deer, man, or only a canoe. The snake does not consider the size or strength of an opponent, but just holds on.

Anacondas kill by constriction, and are non-venomous. They strike first with a bite, and then immediately wrap the victim in their muscular coils. There are some seventy teeth in each jaw. The longest are about 14 millimetres, the thickness of a needle, and angled sharply inwards towards the throat. If any are broken, spares lie flat against the gums ready to assume position. Because of the angle at which the teeth are set, the snake cannot reopen its jaws against the pull of a victim in the opposite direction. The prey is not crushed, but only constricted sufficiently to stop breathing. After death the snake promptly begins to swallow it, all in one piece regardless of size. To perform this feat the anaconda dislocates its jaws, enabling them to open many times their normal size. Simultaneously a slimy coating of saliva is expelled over the victim to lubricate its passage. This gluttonous form of dining has been the demise of many an over-enterprising anaconda. Among those that I found dead over the years were one that had swallowed a full-grown deer and got the antlers stuck in its throat; and another that tried to consume a 7-foot caiman. Anacondas frequently kill caiman, but this snake was only 12 feet long! Sometimes, if the prey is too large, the snake is able to regurgitate it. I lost a hound once, along the Essequibo, and spent the whole day looking for him. Finally he was found, lying dead on a rock bar near the water, covered in a tell-tale coating of saliva.

The majority of anacondas one encounters are less than 15 feet long, under 50 pounds in weight, and the thickness of a man's thigh. They have a life span of twenty years and as they get older their girth and length increases remarkably. Some become the giant specimens about which tales are told. The biggest I ever saw was at the Wai-Wai village of Kanashen. It had been overpowered by twenty Indians in the Kassikaityu River. Despite the manpower it was almost impossible to hold in a stretched-out position for measurement, but the tape read a conservative 26½ feet, and the weight was nearly 400 pounds.

Jimmy Brown and I were taking a tractor one day across the Rupununi River, in water about waist deep. He was driving and I was standing on the back. Suddenly we hit a bump and our wheels bounced over a huge snake that must have been floating downstream. Two coils the thickness of motor-car tyres reared out of the water, but we saw neither head nor tail. This was certainly a specimen of 25 feet plus.

The anaconda is easily captured. Controlling it is the problem. My first move was always to grab them with both hands close behind the head. If the snake was asleep this was not difficult. Unfortunately, at that point they wake up and before you know it your body is a seething mass of snake. If you haven't got anybody to help you, then you're an idiot for being there. I was one, two or three times before I learned. Your only chance is to squeeze it by the throat with one hand, and try and unwrap the coils with the other. You may find this rather difficult, but the anaconda's throat is soft and he doesn't like being choked. If you're lucky he may help you by unravelling himself, but don't count on it. I nearly drowned a couple of times wrestling anacondas—they're just a lot better at grappling than we are.

In January 1968 I received an offer from the producers of "Wild Kingdom", a well-known television series in the United States. Each episode involved a wild-life adventure in different parts of the world dedicated to the conservation of the many rapidly disappearing species. The offer was as co-host to the programme with Marlin Perkins, the famous American zoologist. Apart from the satisfaction of participating in a morally worthwhile project, the saving of the world's imminently threatened fauna, the assignment had a ring of action to it. Many of the episodes would be filmed underwater, with sharks, dolphins and giant rays, while others would be in Africa on horseback with the lasso. The thought of matching pony and lariat against such exciting quarry as giraffe and buffalo, made the "Wild Kingdom" offer one that I simply was unable to resist.

Before I could leave Dadanawa my prime concern had to be the well being of the animals I had collected as pets. I decided to give all my cats to the Georgetown Zoo. At one time I considered returning them to the wild but abandoned this idea for two reasons. First, Chico and Choo-Choo had no fear of humans, and I was concerned that through hunger they might resort to stalking Indians. Secondly, the smaller cats, the ocelots and the margay, had been raised from cubs, and it was unlikely they could successfully hunt and kill their own prey in the forest. In addition, their pelts had become extremely valuable. I felt that some Indians who traded with these skins in Brazil might hear about their release and relentlessly track them down.

Roland went to the zoo because he was a confirmed cattle thief. I would certainly have done the ranch and Roland a great disservice had I released him, for sooner or later one of the vaqueros would have shot him.

Irene kept Shadow, now fully recovered from his battle wounds; and asked to be left the monkeys, and all the smaller animals, like Jill the paca. They loved her, and Irene was like a mother to them all. Tiny and the two young tapirs also stayed. They were a happy group and never strayed far from the ranch house. Every Indian in the country knew the trio and no one would shoot them. They loved to play and swim in the lake behind the house, and were good friends with all the horses. There was an abundance of natural food, but they only had to stand at the garden gate to be sure that someone would hand out the bananas.

The hounds too would remain. They had many friends, for the Wapishana is a great lover of fine hunting dogs. The ranch needed them for predator control, and Mr. Manager and his clan would not have wanted it any other way.

I pondered what to do with Jezebel for several weeks and finally decided to take her with me. There were no quarantine restrictions on birds of prey entering the United States, and I had become so fond of this fierce and magnificent bird that I couldn't bear to leave her.

Transporting her, however, was not going to be easy. She would require a crate large enough to accommodate her for long periods without cramping, yet small enough to fit into the luggage hold of a commercial airliner. Not the least of my problems would be persuading Jezebel to enter the crate.

I finally decided on a plywood box, completely closed in, except for air holes, measuring 48 inches tall, 31 inches wide and 31 inches deep. Being almost dark inside she would stand quietly on a leather-padded perch screwed to the base of the box, and be unalarmed by whatever activities she would otherwise have seen going on about her.

To solve the problem of getting her into the crate, I constructed it so that all six sides were on hinges, and it could be opened out completely flat with the perch in the middle. A week before I was due to leave, I transferred Jezebel from her regular perch to the one on the box, and she quickly became accustomed to it.

Of course, the next question was how would she react to having the sides of the crate closed around her. This would have been simple had I trained her to the hood, but I never had, and it was too late at this stage. The answer was to do it at night, for unlike owls, the eagle's vision is not acute in darkness. A few hours before dawn on the morning of my departure I crept up to her perch with two Indians and quietly lifted the sides of the box into position. At daybreak I visited her and opened the front door about 6 inches. She looked at me fiercely, gave a shrill whistle, but didn't move on her perch or attempt to spread her wings. I offered a tasty portion of fresh meat close to her beak and, though somewhat suspiciously, she took and devoured it readily.

It was a day in March that I will never forget. Jimmy volunteered to take Jezebel's crate to Wichabai in the back of his Land-Rover, and I decided to ride Matador to the airfield to feel his powerful body between my knees for a last time. Irene stood at the garden gate holding Shadow, and he gave a savage growl as I reached out to shake her hand. I trotted down the trail by the dog kennel and Mr. Manager and his pack leapt up at the wire-mesh walls on their hind legs and howled excitedly. I wanted to turn back, but Matador pulled hard at the reins and soon we were cantering across the dry bed of the Rupununi River and along the dusty road to the airstrip.

If the plane had been late that day I probably would have changed my mind. There were things in that country that were good for a man, and I didn't want to leave. But the old DC3 was ahead of schedule and when I rode up she had already landed and taxied into position. The big cargo doors were swung open and the crew began to heave the freight onto the strong backs of the Wapishanas. Sacks of sugar, sacks of flour, rice, salt, guns, ammunition, tools. Jimmy Brown flung the items fiercely into the back of his Land-Rover, just as I had seen him do that first time fifteen years before. Jezebel's crate was picked up and placed on board and my baggage was hurled on after it by a big negro crew member. I climbed up the small metal ladder, and stopped for a moment in the doorway. Jimmy was sitting behind the steering wheel of his Rover, with hat pushed back on his head, applying the finishing touches to his most recently modelled cigarette. I waved, and he acknowledged with a casual flick of the palm, then started the engine. Frantic, laughing Indians scrambled into the back as he pulled off with a jerk, and was soon enveloped in a cloud of dust.

Nothing much had changed since August of 1953. On that day a schoolboy had come. Now it was time for a cowboy to leave the Dadanawa.